INTRODUCTION

'How to become as an Engineer in a Machine shop'.
This book is useful for Engineering students, employees,
employers, HR managers and Interview attending peoples.

I am a Diploma in Mechanical Engineer. Moreover, I was
completed my studies in PSG. Collage of Technology,
which is a very good collage in India. I have 25 years
experience in the Machine shop side.

I have started my carrier in 1982 with Lakshmi Machine
works. First, I am trained in various machines and become
a supervisor. Developed many more items and double
promoted one time in the training period itself. Trained in
most of the areas like Design and manufacturing of Jigs,
Fixtures, Press tools, Re making of machines, Heating coils
and Tool holders. Done Planning, Proving and cost
estimation for the above items.
There I have worked for 10 years and utilized that good
experience wherever I joined. Still I did not satisfied that
experience is not fully utilized anywhere. I never forget
that company in my lifetime. The very good management,
General Manager, Freedom and encouragement. That is
also one of the reasons to write this book.

I0487243

1

'How to become as an Engineer in a Machine shop'.

In this book not only about the machine shop and about some basic principles of the other departments like Sales, Marketing, and stores also we are discussing.

We cannot put a circle and define our work with in the circle and say my work is this much only and I do not bother about the others work. I am paid to do this work only.

We can do teamwork and achieve by putting hands together. More over our growth will depends upon our learning and self-improvements only. We cannot become as a general manager or a manager with out knowing all functions of the company. If you are speasialiced in a machine, you are a good operator for that machine.

However, you know to operate all the machines you will become a fore men. Like that, we know all the functions to become as a manager or a General Manager.

We have to see many peoples are retiring with the same positions by talking about the world wares some thing like that. We have to sharpen our mind whenever it will become blend.

The purpose of this book is to encourage youths. Now the young peoples are having good general knowledge and IQ. They needed some reference to improve their technical knowledge from the experience hands. All experience hands are retiring without recording their technical knowledge to their future generations. In this book, i am sharing purely my 25 years engineering experience in Machine shop, practice that is useful to the Engineering Students, Engineers, and Employers.

For the serial number 3,4 and 5 few details available in this book.

What the system speaks is, we have to record every thing to avoid the secondary mistakes is the concept.

For example, you can write Diary at least from today and at the end of the month you can review then you know the result.

The above systems are available in separate detailed books and please go through it.

We are getting every thing under one roof when we are as a machine operator. More fearless and successful Engineers are created from machines only.

2 Years I am also worked as a machine operator in my training period. After that, I have done lot and lot and switched over many more companies with out fear.

No one company or management will guide you without knowing you. We have to start from the basic as a machine operator even the management refuses because of your qualifications.

The following are the systems most commonly used worldwide.

1. ISO 9000- 2000 system will be common for every Company. We are Engineers definitely know about this system. Please go through this system and implement yourself. In addition, implement even this system is not available in your concern.
2. TS 16949 – Must for all Auto Mobile manufacturing companies. The aim of TS16949 is
   - ➢ Continual improvement
   - ➢ Defect prevention
   - ➢ Reduction in variation and the waste in the supply chain.
3. Total Predictive maintenance
4. Total Quality management
5. Six - Sigma concepts.

This book is written for the Young generations and this will have more details about their responsibilities in a Machine shop. Engineering Students will get an idea to become an Engineer. Select a field and improve their knowledge according to this book. Those who are working in the Engineering fields they should follow this book as a guideline and improve themselves. Use this book as a guideline to interview the peoples al so. Use full to the students to select the field.

All are studying schooling for 12 years and after that doing their selected courses. However, after joining in a firm, do not know what to do. For that, also we can take this book as a guideline to improve your self.

In this book Responsibilities of each field are given. Because in most of the company's employees don't know what he has to do. In addition, an Employer does not know what work he has to expect from the particular person. Therefore, it could be use full to both the Employers and Employee and Students.

Today more and more Job opportunities are there around the World. Every were we can see the 'vacant' boards. Therefore, what the employers are doing is, they are asking the employees to do more than one job and paying for it. That is not a mistake of an Employer. We have to find out the solution for that. We should know the relation ship between the fields. How to combine? What to combine? For that, also we can get idea in this book. This books first correction done by my son R.Arun Kumar and had some innovative thinking and now doing his SSLC.

By
C.K.Rajendra Prasad

BE- A Bachelor of Engineer is recruited as an Engineer.

MBA- Master of Business Administration students are directly joined as a management representative and shortly become as a Manager.

This is a rules fixed by the Manager HR and it will not be correct. The Students are completing their studies around 19 to 25 years and come to work energetically. They are guided properly with very good training and ask to do hard work.

The students after completing their studies, they also work hard and learn more and more. I am giving a simple example now.

As a Machine operator he is doing and getting the following information's
- Drawing study
- Material
- Process study
- Production method
- Mach inability
- Time study
- Tools study
- Problem solving

Therefore, the machine is like an Umbrella where every thing is discussed and experienced in practical. We have no need of wandering and have an unnecessary thinking and worries.

FINAL TIPS

The India is a different country in the World.
An Indian speak many more languages
In Japan, they will speak Japanese
In Germany, they will speak German
In France, they will speak French
In England, they will speak English

Nevertheless, every Engineer should speak only in
Technical language all over the World and there is a no
need of any other languages.

What is the Technical language?
Discussing or telling about the job we can use technical
terms only that is all and it is all geometric symbols like,
Parallelism, perpendicularity, run out, flatness, finish in
microns, etc. And the words in Systems like, PPM, TPM,
TQM, 6sigma, CFT, Keizen,5S, Trend chart, Rout cause
analysis, FMEA, etc.This will give more confidence level
to us and easy to discuss in meeting with any country
peoples.

In many of the companies even it is a reputed concern they
following the some policies that are

DME – A Diploma in mechanical engineer is recruited as a
Supervisor or CNC machine operator.

Contents

SALES ENGINEER

Sales department is the backbone of the company. Where the buyers and sellers and all other areas of business has to meet and make successful. No business can function without professional sales people.

Sales person is responsible for the following activities.

> - Price of the product
> - Service of the product
> - Customer requirements
> - Reliability of the product
> - Price reduction or Economical cost of the product
> - Quality of the product

The all customer requirements are discussed with the concern peoples and reply them. In addition, we have to know the basic principles and function of the product.

<u>Price of the Product</u> The Price fixing of the Product is very important. The sales Engineer will get the customer requirements and estimate the cost with the help of Estimation Engineer. It is important to check the following costs are included or not.

1. Cost of a packing case
2. Rust proof oil
3. Packing materials

Unilateral is the tolerance where variation is permitted in only one direction from the specified dimension

Bilateral is the tolerance where variation is permitted in both directions from the specified dimension

Slope defines the angle of the part.

Counter bore is used to denote the requirement of drilled shape of a part.

Counter sink is used to denote the requirement of drilled shape of a part.

Square is the shape and is used before the basic dimension.

Number of places is indicated that, the same hole or bosses are repeating.

Arc length indicates the length of arc with specified value.

Feature controlled frame indicates specification on a drawing that indicates the type of geometric control for the feature, the tolerance for the control, and the selected datum, if applicable.

S.no	Symbol	Description
27	(F)	Free state
28	(T)	Tangent plane
29	(U)	Unilatral
30	◁	Slope
31	⌴	Counter Bore / Spotface
32	⌵	Counter sink
33	☐	Square (shape)
34	6 X	Number of places
35	⌒35	Arc length
36	⊕ ⌀0.6Ⓜ A B	Feature control frame

Free State modifier symbol is used when the geometric tolerance applies to the feature in its "free state". Or after removal of any forces used in the manufacturing process. With removal of forces the part may distort due to gravity or flexibility or spring back or other release of internal stresses developed during fabrication. The Free State modifier is placed in the tolerance portion of the feature control frame and follows any other modifier.

Tangent plane, when it is desirable to control the surface of a feature by the contacting of high points of the surface.

4. Transport, if it is an export, shipping, or flight charges loaded on a container
5. For export If we are keeping in the wear house. Then add wear house charge
6. Expected Rejection charge
7. Tax to be paid

Service of the product
A customer is buying a product because of that product's features and benefits along with their emotional attachment or feeling about the product. These can be slightly influenced by the salesperson.
The sales person also can influence the relation ship with the customer by keeping the timely delivery and new methodology introduced to do his product to improve his quality.

Customer requirements This is the first important step of the sales. Listening and asking intelligent and interesting questions that are open-ended, gather information, and discuss with the Estimation engineer and Shop Engineers. A good sales person is a listener. They ask plenty of questions, and make notes of the answers. These notes will help them to find a suitable service for the potential customer. A successful sale is when the customer agrees with that solution.

Reliability of the product Sales engineer has to monitor the reliability of the product in the market. What is the trend now and how long it will stable. Who are they competitors for us? How they are marketing and the price competition with them are important.

Economical cost and Profit of the product Many sales people worry about the price of their goods compared to the competition with the market price. It is only when a sales person, or a third party introduces doubt into the mind of the customer, that price becomes an issue. Before that, we can estimate the right price, find out the cost cutting analysis with the machine shop peoples, and ready for the competition.

Quality of the Product The sales person should know the important features of our product. We have to explain how best our product compares to competitors like material and performance is an added advantage. Say for example we are using imported bearings, Taper roller bearing, double row ball bearing, High rpm, Rust proof material etc.

Type of buyers The type of buyers is classified as two types. One is direct and another one is Agents. Buyers are our Customers and we cannot feel the much more differences with them. Treat as same as possible. The direct buyer comes once or twice but he will communicate the quality of our product to other peoples. Who will then come to us? Agents will come often to us. They are encouraged by giving incentives to them.

and datum. Basic dimensions are indicated by rectangular around dimension or not tolerenced directly or by default.

21. Reference dimension is used for information purpose. A reference dimension is a repeat of a dimension or is derived from calculations for information purpose, usually without tolerance.

22. Dimension origin used to indicate the origin and direction of dimension between two features.

23. Datum feature is the feature of a part that is used establishes a datum.

24. All around profile will define the requirement of total profile of a part.

25. Conical taper defines the taper of a hole and angles.

26. Depth is used where the blind holes are coming in a part or how deep its required.

rotating shafts; can be checked between centers by using a dial with a stand.

16. At maximum material condition in which feature of size contains the maximum amount of material with in the stated limit of size. For example, the lower limit of the hole is the minimum hole diameter. The upper limit of the shaft is the maximum shaft diameter.

17. At least material condition in which feature of size contains the least amount of material with in the stated limits of size. For example, upper limit or maximum hole diameter and lower limit or minimum shaft diameter.

18. Projected tolerance zone is used to see variation in perpendicularity of threaded holes could cause fasteners such as screws and studs to interfere with mating parts. So if the projected zone height is given as 15mm it can be checked from its true position projecting 15 mm height from threaded hole. Possible to measure by CMM only.

19. Regardless of feature size is used to indicate that a geometric tolerance or datum reference applies at any increment of size of the feature with in its tolerance limits.

20. Basic size is a numerical value used to describe the theoretically exact size, orientatation, location

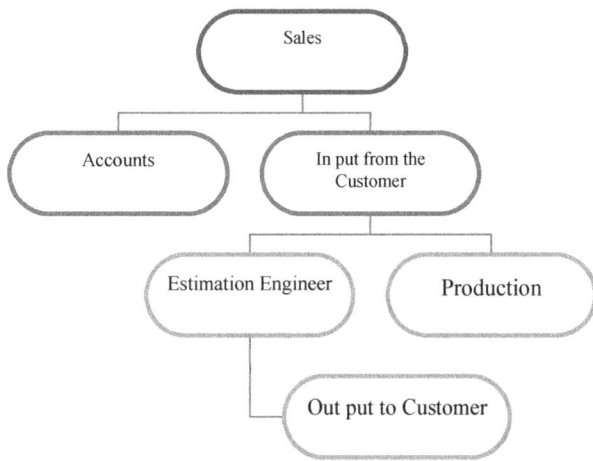

1. The incentives of Sales Engineer is given to increase their total number of sales

2. The incentives of Sales Engineer is given to increase their total turn over of the company.

The marketing will also be a Planning process. The long term and short term forecasting will be required to plan. In the marketing, we can plan monthly and yearly depending upon the product.

The Marketing engineer is responsible for the following steps.

> Get through knowledge of the product.
> Search the similar products in the market.
> Price of the similar product and its quality.
> Quatityable movement of the product.
> Marketing techniques
> Ask their end users for improvements.
> Sales of the product
> Turn over of the company

Get through knowledge of the product.

The Marketing engineer should know the important features of our product. We have to explain how best our product compares to competitors like material and performance is an added advantage.

Search the similar products in the market.

Why we search the similar product in the market?

Un Fort natively the similar product is available in the market we can do lot of researches to manufacture and market the product. If the product is not available and we are launching the first time it is easy to market and make profit.

Price of the similar product and its quality.

Before launching or before manufacturing our product, similar product cost and quality must be known and we can take it as a input for our Sales and marketing. We have to

13. Radius is asked like R5, R10 etc. The 'R' stands for Radius and next follows a numerical number, which is its value. Can be measured by male & female radius gauges.

S.no	Symbol	Description
14	⚹	Circular run out
15	⚹⚹	Total run out
16	Ⓜ	At max. material condition
17	Ⓛ	At Least material condition
18	NONE	Regardles of feature size
19	Ⓟ	Projected tolerence zone
20	20	Basic dimension
21	(21)	Reference dimension
22	⊕—	Dimension origin
23	-A-	Datum feature
24	⚲	All around profile
25	▷	Conical taper
26	↧	Depth / Deep

14. Circular run out is the term used where the circular or cylindrical parts has required to run is in range or in out of range or with in the said limit. Usually given for rotating shafts; can be checked between centers by using a dial with a stand.

15. Total run out is the term used where the circular or cylindrical parts with number of steps has required to run is in range or in out of range or with in the said limit. Usually given for stepped

7. Angularity is its meaning it self denotes the angle of a surface, which is measured by a Bevel protractor.
8. Perpendicularity is a term that, how perpendicular the hole, bore or any projections related to a datum face or a base. We can measure by holding the work piece horizontally with the use of L-Bracket at surface table by Dial with height gauge.
9. Parallelism is its meaning itself to measure its Parallelism of the parts from its bottom to top surface is measured by just placing the work piece on the table and use feeler gauges and see whether it is entering or not. We can use Dial indicators al so to measure.
10. Positioning symbol is given normally in Bolt circle diameter with reference to the center bore as a Datum. Take 1/3 of the tolerance given. For example +/- 0.3 mm is given means, measure with in +/- 0.1 mm . Position is theoretically exact location of a feature established by basic dimension.
11. Concentricity is that the number of circles formed at the center as an origin to be concentric to each other. Normally checked one bore to another bore, which is in the same centerline. Use stepped plug, Trimos, CMM etc to measure.
12. Symetricity is the term that says both are identical or it could be a mirror image used in profiles where the shapes are dimensionally identical or mirror.

Compare the prices, quality and cost cutting possibilities to competitive with them are give idea to us to start or with draw the project.

Quatityable movement of the product.
The quantityable movement of the similar product is much more important. Say for example already the similar product is launched in the market and its selling quantity is 5 machines per month at an average. In addition, he is making the profit hand to mouth will be reconsidered or the project could be dropped and go for some other project.

Marketing techniques
To market the product we follow the following steps.

➢ Advertisement through channels and Yellow pages.
➢ Advertisement through Web sites.
➢ Thro' agents with better commissions.
➢ Participating in the National and International exhibitions.
➢ Conducting seminars in various places.
➢ Direct marketing with Sales personals etc.

Ask their end users for improvements.
If we find the similar product and search their customers and ask so many questions about the products such as, advantages and the disadvantages of the product.
Implement the improvements before the first launch itself.

In most organizations like Job order and product manufacturing companies, we will have yearly plan only. That is like a short term fore casting.

The customer or a buyer will not tell them immediately. They will provide contract for a year and in-between they will tell, we do not want the product and the season is become dull. You will supply only when we will intimate you or the stock is enough for three months we do not want for three months. What will happen? The three months will become idle. We can go for the new product is not so easy. We all ways develop a new product side by side and out source it and keep ready.

In the machine manufacturing company's we can do long term fore castings.

For example, we can take a Lathe manufacturer. In lathe manufacturing, types of lathes available are 1.Center lathe 2. Turret lathes 3. Copying lathe and CNC. Lathe.

For 1, 2 and 3 Headstock, tailstock and bed we can make common is possible. With small changes, we can achieve the target level.

A few Companies' may look at a practical plan, which stretches three or more years ahead. To be most effective, the plan has to be usually in written format.

S.no	Symbol	Description
1	—	Straightness
2	▱	Flatness
3	○	Circularity
4	⌭	Cylindricity
5	⌒	Profile of a line
6	◠	Profile of a Radius
7	∠	Angularity
8	⊥	Perpendicularity
9	//	Parallalism
10	⊕	Position
11	◎	Concentricity
12	≡	Symetricity
13	R	Radius

4. Cylindricity is to see the cylindrical part is oval or circular with in the limit. The allowable oval is 1/3 of the bore tolerances and is measured in shafts by a Micrometer.

5. Profile of a line is noted that it is not the exact profile can be produced its vicissitudes with the allowable limit and is measured in Profiled shaped Work pieces.

6. Profile of a Radius is noted that it is not the exact profile can be produced its vicissitudes with the allowable limit and is measured in Radius shapes wherever in Work pieces. It will come in bores and outer of the parts as it is or while machining and is measured by radius gauge or templates.

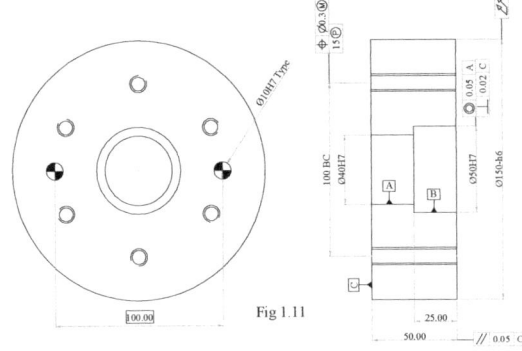

Fig 1.11

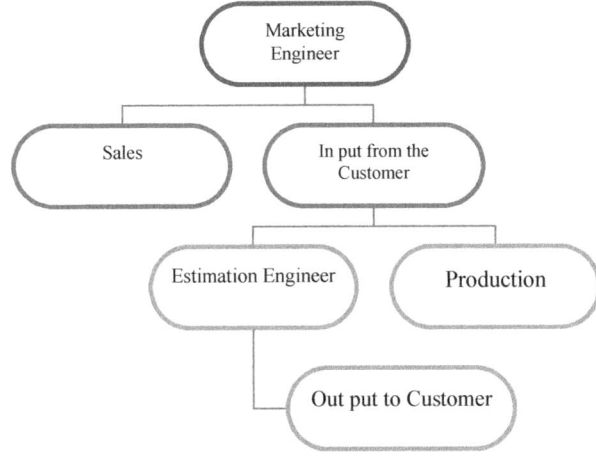

The above figure 1.11 shows some of the important geometric symbols normally used.

The following table shows the most commonly used geometric symbols

1. Straightness its meaning itself to measure the straightness of the parts used like, Length flats, L-Angles, Beams, Pipes, and Rods etc. These are the items can be purchased and used directly with little machining process in which the straightness value is required.

2. Flatness its meaning itself to measure the flatness of the parts is measured by hanging the Work piece on three points and making it zero, see over all.

3. Circularity is to see the bore is oval or circular with in the limit. The allowable oval is 1/3 of the bore tolerances.

The Difference between the Marketing and the sales is little bit only. Both are involved directly with the customers. The marketing people will somewhat headache than the sales person.

The marketing strategy needed to achieve marketing objectives. To be most effective, objectives should be capable of measurement and therefore 'quantifiable'. This measurement may be in terms of sales volume, money value, market share, percentage penetration of distribution outlets.

New Product could be developed with improvements not completely changing the older one. Then only we can get more money in the spares. The Spare parts are the very important thing and will give more profit. The market trend is depends on the availability of the spares of the any product. If it is necessary to go for a new one then its ok and keep ready your spares for older once in the market for some what time..

As for as the engineering field is concern, Change in design and improvements to the customer requirement will be enough to survey in the market.

The price will be decided to different places with their cost of living or money market in those places. For example, money market will defer place to place or country to country.

Purchasing refers to a function in business whereby the enterprise obtains the inputs for what it produces, as well as other goods and services it requires.

<u>Responsibilities of Purchase Engineer</u>

> ➢ Meet Customer requirements
> ➢ Price of the product or element
> ➢ Reliability of the product
> ➢ Timely delivery
> ➢ Life of the product
> ➢ Quality of the product

CNC milling cost per hour – 500 rupees
Therefore, we need 5 times of the productivity in CNC than ordinary milling.
We cannot simply fix the prices for the work piece and consider all the factors.

For the grinding process, every body knows its finish. However, various grinding machines are available.

1. Surface grinding machine
2. Cylindrical grinding machine
3. Bore Grinding machine
4. CNC profile Grinding machine
5. 0.09 Stone Grinding machine
6. Belt Grinding machine
7. Hand Grinding machine

So the machine finish will defer machine to machine. In addition, the finish is purely depending on the speed, Feed, and type of tool we are using.
For example, 12.5 micron finish rough milling is enough. We can give more depth of cut, less speed and more feed to save the time, which is enough to achieve the 12.5 microns roughness value. At the same time to obtain 0.8 finishes, by changing the speed feed and adding no. of passes is achievable.

We can see why in the above machining symbols the processes are repeating. We can see the surface finish 0.8 microns to 12.5 microns milling process is given. The finish can be achieved in any method with cost effectiveness is the main point. For examples,

The milling, Drilling or Turning will give 0.8 microns finish.

The same we can achieve in CNC and grinding process too. However, it is costly.

If the process will contain critical geometric tolerances with the same finish then we can go for CNC. Then the machining cost will become more.

If the ordinary milling will produce 2 pieces at the same time the CNC will produce 8 pieces will be the profitable.

Advantages of CNC machines
  ➤ More Productivity
  ➤ Fewer work forces
  ➤ Consistency in Quality
  ➤ Less human dependability

Disadvantages of CNC machines
  ➤ More Tool cost
  ➤ More Insert cost
  ➤ Skilled peoples
  ➤ More capital investment
  ➤ Continual Quality check
  ➤ Cost of the machine etc.
The milling cost per hour – 100 rupees

Meet Customer requirements The Customer for the Purchase engineer is the peoples whom they are giving inputs internally from the company itself. We have to meet the specifications given by them to the exact requirements.

Price of the product or element The cost of the product is compared with the number of available sources before buying it on the quatityable basis. Say for example, we require 1000 numbers of M10 Allen head screw per month. This is a Quatityable business. We are getting 12 rupees per piece in Hardware stores.

But we talk to a manufacturer or a dealer and buy for 11 rupees then profit will comes around 1000 rupees per month. If we will have 100 items like this then the profit will be enormous. It will become use full when we will go for the price reduction in our product.

Reliability of the product We have to find out the reliable sources to purchasing the items. Every time goes on to changing the sources will get problem in future. If we found some other sources in reduction in price, do not buy immediately. We have to analyze them, is they are reliable for long-term business.

Timely delivery The timely delivery in case of the assembly, mechanical maintenance, and electrical maintenance will reduce the machine idle time and delay in delivery of the product. Not only the above items, all the items directly or indirectly related with our supply of our product commitment.

<u>Life of the product</u> The life of the product is essential in case of Hardware items like paints, rust proof oils, resins, bolt nuts, screws, grinding wheels, abrasives, welding rods and adhesives. We have to purchase those items by showing the expiry date and check is it falling before our usable date. Bolt nut and screw there is no expiry date but we will check with the assemblypersons or refer the customer complaints to buy the same company regularly.

<u>Quality of the product</u> As well as the cost, life, and time, the quality of the product is much more important. For example we brought M10 Allen head screw with cheaper rate in right time near by and the fitter is assembling the product. While assembling it, its head is damaged and not tightening, it will not solve the purpose.

The Purchase Engineer will get input from the following departments.

> Planning Department
> Production Department
> Quality control Department
> Industrial Engineering Department
> Mechanical maintenance department
> Electrical maintenance department

<u>Input from planning</u> There are 2 type of purchase one is consumable goods and another one is tools. The planning department knows the standard items which are required to manufacture our product. He will consolidate the standard

MACHINING SYMBOLS

Machining symbols – Machining Symbols are the impotent factor, is most commonly used world wide in every drawings. Before going to decide the process or to design a Jig or a Fixture, careful study of machining symbol and Geometric symbol is necessary.

The following table shows the most commonly used machining symbols

S.no	Roughness Grade Symbols	Roughness Grade Symbols in use	Achived by process
1	N12	50	Gas cutt. & Band saw cutt.
2	N11	25	Forging
3	N10	12.5	Rough Milling
4	N9	6.3	Milling, Drilling
5	N8	3.2	Milling, Reaming, Turning ,
6	N7	1.6	Boring, Planning, Shaping
7	N6	0.8	Broaching, Grinding
8	N5	0.4	Grinding, CNC milling, CNC turning
9	N4	0.2	Grinding, CNC milling, CNC turning
10	N3	0.1	Honning
11	N2	0.05	Honning
12	N1	0.025	Lapping

Others – This facilities are not mandatory can be provided by big organizations or a good profit centers only. Those are Bonus, Incentives, Tour allowance, Sweets, and snacks at the time of festivals.

Environments

Working environment or the atmosphere is very important. For example, one Employer can say, we have given all the above facilities with better salary but still peoples resigning more. Why it is happening?

That is because of the working atmosphere. They will do painting, gas cutting, grinding, straightening, Rust proofing oil spray, and shouting everywhere inside the company. It has illiterate the operator.

Most of the company has started with out planning; they will start for their amount in their hand.

They will make a shed and thro' the machines and operators inside the shed. Make facilities one by one. They will not full fill the requirements immediately but give higher salary than the limited company and get failure in Labor management. Then permanently put a wanted board in his gate.

To day's trend, growth will be there but organization structure will not change drastically. So start a company as per ISO certification level, even its 2 machine or 3 machines. When company is, growing working atmosphere will also grow simultaneously and stop resignations.

items and give it to purchase as a monthly requirement before one week of the month.

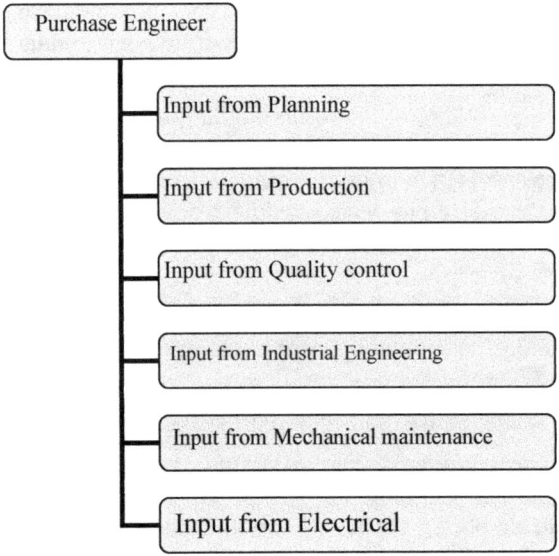

The consumable goods are the Bolts, nuts, screws, paints, emery cloths, raw materials, etc.

One more item is the stationary item, which can be raised by the accounts department. Those who will require their stationeries will be informed thro' a written format to accounts. The items like note bboks, pens, pencils, rubber, floppy disc, CD writers, printer ink, ribbons, and Papers etc.

Input from production Tooling items like Drill, reamer, end mill cutter, slot mill cutter, face mill cutter, micro boring bars, inserts, grinding wheels etc. Initially this input will come from planning department and production department will monitor the monthly consumption and raise the indent.

Input from Quality The quality requirements like Instruments Vernier caliper, Micrometer, Plainplug gauge, Thread plug gauge, Dial gauge, bore dial, height gauge etc. The instruments will not required often and frequent requirement will be the plug gauges only.

Input from Industrial Engineer The industrial engineer is called an s a Development engineer in some companies. He will raise the indents for capital goods like Tool holders, machineries, computers, table, chairs, machine spares etc. What is my 'indent' it is a format most commonly used in many of the industries. In which we will fill the Specification of the required item, Qty, purpose of the item, in which date it will be required, raised by, and approved by etc.

Input from Mechanical maintenance They require gears, Bearings, worn out parts of the machines and hardware items.

Safety – We have given the appointment to an operator with PF, ESI, good Toilet, Food facilities with leave. Now he is coming to work with satisfactory.

However, he met an accident inside the factory because of a safe guard not provided in his machine. This is a very big loss for that operator and as well as to the factory.

Factory HR again goes to search for a good operator will make expenses for advertising, time loss and production loss will be the result, up to that operator's recovery.

Safety equipments like Goggles, Hand glues, Shoes, Uniforms etc. should be provided.

Is it the uniform is the safety equipment? May be, give the uniform, and ask each operator to come with tuck in condition.

Because in the machine we cannot provide safeguard everywhere. For example, chuck and a lead screw in a lathe. The operator shirt will roll in to the lead screw in a lathe. Not only in the lathe, in every machine there is a loophole for rolling the shirt.

Recreation – Not only for the operators and for all the employees we have to give some refreshment. For example, provide small things like a TV in a lunchroom. Provide a small space to play a shuttle cork. Ask employees and their family to participate in Music, dance, Sports once in a year on Independence Day and give small prizes like gift articles and good books will give good result to the company.

gratuity to his employees, on the discontinuance of service, if they have served a minimum of 5 years of continuous service (except in the case ofdeath).

Toilet – One can ask, is there any company with out Toilet?. That is not a matter and it could be kept clean and neatly. Most of the diseases starting from the toilet only.

Diseases like Fever and Malaria will come easily. That is the very big indirect loss for the Company. He will take leave often and realize and switch over to some other company.

Canteen – For example in a Company Labors are taking more leaves and leaving every month with out information. What will be the rout cause analysis? Go to Human Resources department get fast 6 months resigned peoples address and find how far away they are coming. If the distance is more he will find difficulty to get food. He is a bachelor and has some other nativity and taking room nearer to the company. He doesn't know to cook or didn't find enough time to cook or cooking items are not available near by. So we provide canteen and enjoy the benefits both.

Leave – There is an earn leave and sick leave is mandatory to give all the employees. Leave is essential for every body and don't refuse every time. The essential leaves like marriage function, Illness and their relatives' death etc. The turn over is calculated with considering these leaves also or provide alternate arrangement by considering this.

Input from Electrical maintenance They require electrical items like motors, wires, switches, limit switches, cables, capacitors, lights, fans, regulators, PCB, extension boxes, etc.

The difference for consumable and Tools are much difference. For consumable items, we recruit Bachelor in science people and for tool purchase, we can recruit the production engineers with 1 or 2 years experience. This will solve the purchasing of correct items. If the company is small, then the single person who knows better about tooling looks both after.

Estimation engineer

The roll of and Estimation Engineer is very very important for the company. Because he is the Key person to the Company, the cost and profit is purely based on his Quote only. Most probably, we can select such type of person from development side only.

1. He first knows to study the Drawings

2. He knows the metallurgy like, materials and its standards

3. He should know how to process the elements

4. He should know what the machines to be involved are

5. He should know what will be the time required to process

After doing the above steps, he has to quote to the customer

1. Drawing study

What will be an Estimation Engineer can Study in the Engineering drawings.
It could be purely a science and Mathematics'

➢ See the length, width, and thickness of the work piece

➢ Then see whether it is possible to accommodate in our machine.

➢ See the critical parameters like parallelism, perpendicularity, circularity, concentricity and finish is achievable or not

➢ Check the tolerances are achievable or not

➢ Whether any fixture is required or not

➢ Whether any Gauges are required or not
➢ Whether any special instruments are required

➢ Qty of an order

2. Metallurgy

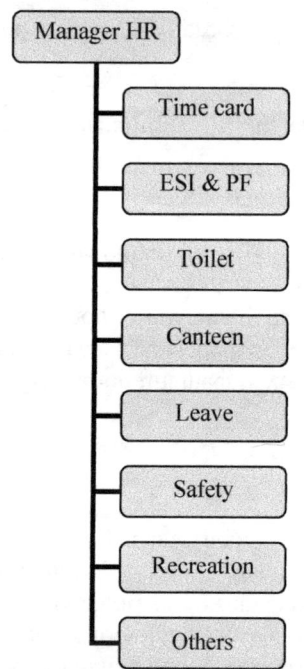

Gratuity - It is a mandatory tax-qualified, defined benefit plan paying ½ months salary for every year of service / work completed in lump sum at retirement. Every employer employing more than 10 persons must pay

and the employee and the same is payable in lump sum on retirement. The current rate of PF contribution by a member is 12% of Salary / Wages (Basic + Dearness Allowance) with matching contribution from employer. Settlement can be done only after a waiting period of two months from the date of resignation but in cases of members leaving abroad, settlements can be done immediately and settlements are immediate in case of female members who resign from the services for the purpose of getting married.

➤ In and out of the material details like material standards and its composition are not possible. Because it is a big subject.

➤ However, should know general material properties like hardness, specific gravity, Weight, and cost of the material.

➤ Then only cost of insert can be calculated. In general, the materials like OHNS, HCHCR, HDS, HSS, SS, CF8, and ALL STEEL CASTINGS are harder than the cast iron materials.

➤ For cast iron materials, there is no much need for adding up of an insert cost. However, for Steel items adding up of an insert cost is must.

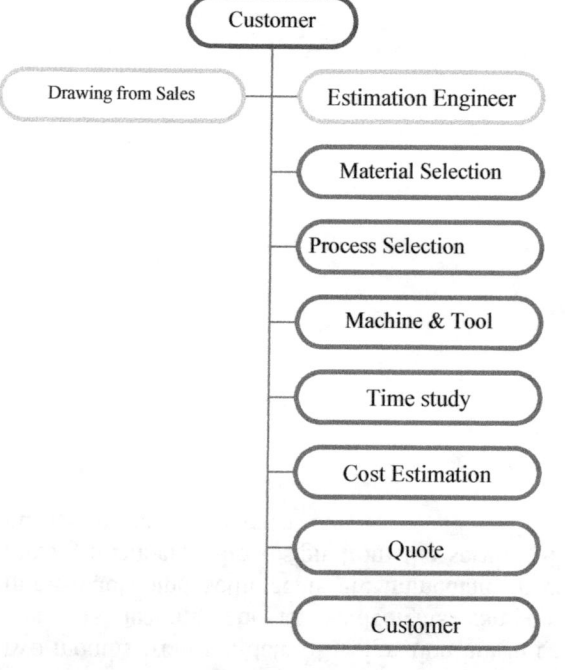

All non seasonal factories using power and employing 10 or more persons and non-power using factories employing 20 or more persons are coverable under the ESI Act. All employees in such factories getting wage up to Rs.6500/-per month. are coverable under the Act.

Benefits under the Act:

The following benefits are provided to the insured persons and their family members under the provisions of ESI Act.

1) Medical Benefit.

2) Sickness Benefit

3) Maternity Benefit

4) Disablement Benefit.

5) Dependant's Benefit.

6) Funeral benefit.

The First benefit i.e. medical Benefit is provided by the State Government through the ESI Scheme and the remaining five benefits, which are cash benefits are provided by the ESI Corporation directly.

The ESI is the very good facility provided by the government and will provide more security to the entire family. In present condition minimum 500 rupees for medical expenses is required to entire family.

Provident fund (PF) - It is a mandatory, tax-qualified, defined contribution, retrial benefit plan wherein equal contribution at the rate of 12% is made by the employer

We have to take this book as a reference and interview al so will give good results.

The most important things to be watched while interviewing the person is his discipline regarding

➢ Dressing
➢ Manner of answering,
➢ Sitting position,
➢ Suitably answering or not,
➢ Depth of his voice
➢ Discipline level at the previous employment too.

Time card – Even in Computerized factories, they introduced as a pay cards.
They will have a in – out punching card, which is for his attendance purpose only.
In Pay card information's like Drawing number, Description, Job loading and un loading time, Completion time and Idle time will be there. After ending of the each shift, he will fill it up and get the signature from the shift Supervisor.
Then it could be entered in to the computer and based on that he will get the salary. This will helps to avoid time fluctuation between one operator to another.
In small Factories Time card have information's like Ist shift, IInd, IIIrd shift and over time only will be there. The Supervisor can sign while the operator entering and while going out.

Coverage under the ESI Act.

3. Machining Process

First, we have to draw a flow diagram.
What is flow diagram?

It will define the flow of material movement from one machine to another machine with a continuous flow without breaking it.

The continuous flow is very important; the uneven flow will break the profit. I am explaining a simple flow diagram with example. Which one is a stepped shaft with a slot and a hole as above in the drawing.

In the below flow diagram shown the inspection to each operation. Because each and every operation are carried out and getting ok will move on to next stage. The below flow diagram explaining very simply all the details of the machining process.

As per the flow diagram what it will speaks, First step is the raw material and once raw material is received we can start 1st operation as turning and inspect it. If found ok go to second operation as turning second side. If it is found ok, go to third operation as slot milling. If it is, found ok then move to third operation as drilling, inspect, and do debarring.

Final operation is Final inspection, where all the final requirements are checked thoroughly before dispatch.

In every operation inspection is given. That is compulsory to inspect at every stages. If the work piece is, get rejected in first or second operation. We will save the time at least in further operations and other operational cost.

The Inspection is carried out at each every stages by the inspector or by the operator it self depends on the confidence on the work piece and its type and value.

Machine requirements

> First operation is turning first side

For Turning, what are the machines are available?

Lathe? If it is a lathe means is it is a Center lathe or CNC lathe and it is a Vertical or Horizontal. Turning second sides, also decide similarly. Check fixture required or not.

> Third operation is slot milling

A milling machine required. If it is a Vertical milling or Horizontal milling or vertical machining center or horizontal machining center or only a slot milling machine? Check fixture required or not.

> Give ESI benefits
> Give Provident fund (PF)
> Give good and clean Toilet
> Provide food facilities (Canteen)
> Give earn Leave and sick leave
> Provide Safety environment
> Provide Recreation clubs.

In many of the companies, the manager HR is not interviewing the peoples. They are asking the concern person to interview and select to their requirements. This is not correct and what the interview committee person doing is they will not select on the skill basis.

They will select un skilled people, who will not damage his position and will not over take him also. The result will become poor and the un skilled one will go in short time. The net result is resignations will become often and headache to the company.

I have attended hundreds of interviews with reputed concerns and one favorite question will be asked is "Tell me about your self"

Then he will tell some thing and based on his answer the HR will ask some thing and the interview will become an end.

This is not a right approach and one small suggestion is; Bee with the person who is interviewing person up to the last minute will give solution.

We have discussed earlier chapters about the many cost saving and promotional activities.

The above activities are purely on the hands of the General Manager. This will increase the turn over and save the cost directly or indirectly to the company.

> The General Manager judgments and recommendations to the Management are based on his own skill.
> The General Manager will act like a politician while reporting to management.
> He never project any week ness of his subordinates
> Do not give loose talks about his employees.
> The Customer complaints and rejections faced will not intimate immediately.
> He has to tackle his subordinates as well as the management.
> Those who are not supporting to him and even he is a management recommended person, send him out is the first step of the success.
> The persons who are sending out will be, with warning or changing his profession or by de promoting him and not by simply asking to go out.

MANAGED HUMAN RESOURCES

The main responsibility of the Human resource manager (HR) is to provide or recruit skilled peoples to the company. He is responsible to the following activities.

> Provide the Time card

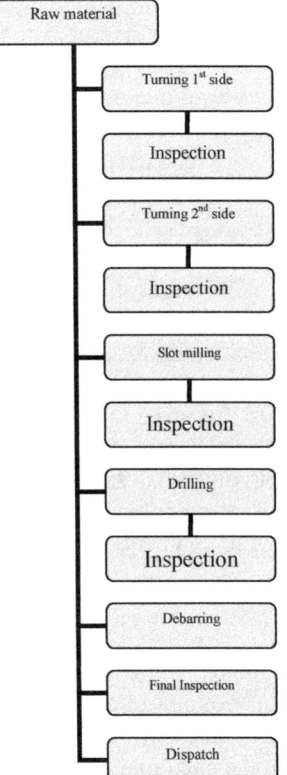

> Forth operation is Drilling

A Drilling machine is required. If it is a radial drilling or Bench drilling or Column drilling machine or simply a hand-drilling machine is enough. Check drill jig required or not.

> Fifth operation is a Debarring operation

Debarring is a process to avoid sharp edges that will not harm our hands while handling the work pieces after finishing it. So what will be required? Is it a Hand grinder or Debarring cutter or Emery sheet or Oilstone or Files or both? Mostly both wills required

> Sixth operation is Final Inspection
After completion of a job, it could be inspected thoroughly in every aspect. Standard instruments like Micrometer, Vernier, Height gauge, Dial gauge are available in a factory. Its costs are not added but special instruments like CMM, Trimos and Special gauges particularly made for this to be added with machining cost. Check Gauges required or not.

> Seventh operation is a Dispatch

Is it a dispatch cost is required? 100% required otherwise your profit would not come. What are all comes under dispatch?
1. Cost of a packing case
2. Rust proof oil

Manager has to implement the above Systems which ever required to his company will have a power and management support too. He has to motivate peoples by giving some rewards and promotions for cost saving activities will give better result. Today's implementation is tomorrow's fruit for us.

Turn over
The Turn over is not simply a Qty or total money produced. Simply for example our turn over is one crore and the profit is one lakes excluding the expenses incurred by the company is no use. The motivation given by the management also in profit basis will give high growth to the company.
To achieve the turn over we implement the following steps.
> By introducing the Incentive schemes
> By giving the Profit shares
> By giving Promotions
> By giving training
> By recruiting skilled peoples
> By motivating Keizen activities
> Down time of the break down maintenance
> Splitting bonus into two intervals

The bonus will be a great issue, and it will be given 60% before Deepavali, and balance before Pongal will give good results.
In most of the Companies, the resignations are based on the Deepavali functions only and it is true. There fore splitting it, it will give better result.

What ever it may be the Product manufacturing or Machine manufacturing or Job order company, the above method is applicable.

Customer complaints

The Customer complaint is the most important factor to consider. The farmer Tamil Nadu chief minister Mr. M.G.Ramachandran will ask the price of the rice every day and control it. The rice is the basic need of his customer (voters). Therefore, up to his death; He is only the winning person. Like that, the Customer complaint is discussed daily on the meeting and resolved it.

System implementation

With out implementing any system we will not survive and the pressure will rise for every body. The following systems are available. For example our body is high systematic than ever and any one of the system fails in our body we will become sick. If we will get treatment in the initial step
It will avoid the serious illness.

> ISO 9000- 2000 system will be common for every Company.
> TS 16949 – Must for all Auto Mobile manufacturing companies.
> Total Predictive maintenance
> Total Quality management
> Six - Sigma concepts.

What the system speaks is, we have to record every thing to avoid the secondary mistakes is the concept. The General

3. Packing materials
4. Transport, if it is an export, shipping, or flight charges loaded on a container
5. For export If we are keeping in the wear house. Then add wear house charge
6. Expected Rejection charge
7. Tax to be paid

With all the above an Estimation engineer know how to calculate a machining time is very important. The simple calculation is discussed as below.

First, we must know the machining formulas

V- Is the cutting speed (meters/min)
D- Diameter (mm)
N- Rpm (Revolution per minute)
F – Table feed (mm)
, f – feed per tooth (mm per tooth)
Z – no. of tooth (of a cutter)
L – Length of cut
T – Machining time   Ref fig 1.1

Turning – 1st side by CNC Lathe
Raw material diameter is 54 mm
First side turning length is 150 mm
Facing width is 5 mm per side & length is radius of the work piece

$$V = \pi D N \frac{}{1000}$$

$$200 = 3.14 X 54 X\ N/1000$$
$$N = 200000/3.14 X 54$$
$$= 1179\ rpm$$

Machining Time (Turn) $T = L / F\ X\ N$
$$= 150 / 0.2\ X\ 1179$$
$$= 0.636\ min$$
$$= 38.16\ seconds$$

With in a single pass it is not possible we require at least 3 passes

1. So machining time = 38.16 X 3 = 1.9 min

Machining Time (facing) $T = L / F\ X\ N$
$$= 27 / 0.2\ X\ 1179$$
$$= 0.115\ min$$
$$= 6.9\ seconds$$

With in a single pass it is not possible we require at least 3 passes

2. So machining time = 6.9 X 3 = 0.35 min

Turning – 2nd side by CNC Lathe

Machining Time (Turn) $T = L / F\ X\ N$
$$= 50 / 0.2\ X\ 1179$$
$$= 0.212\ min$$
$$= 12.72\ seconds$$

conduct daily meeting and the following points are required to discuss during the meeting.

➢ Name of the product
➢ Purchase order number and date
➢ Delivery Date
➢ Special instructions if any given, that requirements
➢ Raw material specification
➢ Drawing number
➢ Sequence of operations
➢ Machine number
➢ Requirement of jigs or Fixture

The above features are tabulated and give the space for tick the column or row.

➢ First, we check the Product requirements and its specification and go to next step as the process requirements.
➢ In machining process, today the work piece where its stands will be ticked
➢ Is there any jig or Fixture required or under repair and special tool required for further process will be intimated to the concern people in the meeting itself.
➢ The problems and its route cause analysis to be discussed and solved in the meeting itself.
➢ The purchase items listed separately. The same is given to purchase with required quantity to procure before usage.
➢ If there is, any Mechanical or Electrical problems will be there to be sorted out.
➢ If there is an assumption regarding design and Drawing to be sorted out.

the marketing, we can plan monthly and yearly depending upon the product.

Marketing techniques

To market the product we need the following steps.

- Advertisement through channels and Yellow pages.
- Advertisement through Web sites.
- Thro' agents with better commissions.
- Participating in the National and International exhibitions.
- Conducting seminars in various places.
- Direct marketing with Sales personals etc.

The General Manager has to visit with the sales and Marketing personal frequently to the Customer end and implement the above marketing techniques to improve sales.

Production Planning

The General Manager knows well about the sales of the year and future. Based on the yearly plan he has to conduct meeting with all related department heads.

We have to make yearly plan with Sales and Marketing personals itself and among the yearly plan, we give priority month wise and routed through sales to planning.
We can plan the production neither by the Customer basis nor by the turn over basis. In turn over basis, allowing instant customers under no load condition is ok. Preference might be given to the Customer basis is good. We have to

With in a single pass it is not possible we require at least 3 passes

3. So machining time = 12.72 X 3 = 0.64 min

Machining Time (facing) $T = L / F X N$
$$= 27 / 0.2 X 1179$$
$$= 0.115 \text{ min}$$
$$= 6.9 \text{ seconds}$$

With in a single pass it is not possible we require at least three passes

4. So machining time = 6.9 X 3 = 0.35 min

Total machining time = 1 + 2 + 3 + 4
$$= 1.9 + 0.35 + 0.64 + 0.35$$
$$= 3.24 \text{ min}$$

For chamfering, we have to add 10 seconds

Next, we have to go for idle time.

What is mean by idle time? It is simply a non-cutting time occurred during the machining operations.

What are the operations involved in turning? (Fig 1.1)

OD Turning, facing, turning, Finish turning, face finishing and chamfering

Very how much tool we are using?

For the all above operations, only a single tool is enough. Therefore, the tool approach time we have to add as 10 seconds its depending upon the machine. What they have specified we have to add or 10 seconds.

In addition, we have to see opening and closing of the doors. Clamping and de clamping of the work piece. In addition, insert change time.

Normally the table or slide will move five to times higher than the cutting time. So 5% to 10 % of the machining time will become as idle time

In turning tool will return ideally for next cutting operation and some times we index the tool head 4 to 6 times if we use 4 to 6 tools.

In VMC or HMC, also the same will happen. Most of the company is not considering this idle time. Why i am forcing to see the idle time? It will affect your profit directly. Because if your or doing the bulk production it will affect enormously, we can see with above example also.
Next is the internal transport. We can move work pieces one machine to another machine. Even if you are doing in the automatic system, loss in seconds will come

Therefore, Tool movement, Clamping time, opening, and closing, tool return, insert change, operator fatigue all will come under idle time. At that time power loss, machine hour loss, operator salary loss will happens.

The origin for the company is sales, so his first work will start from sales. In sales, the General Manager (GM) monitors the following work.

Price of the Product The Price fixing of the Product is very important. The GM. decides the final cost of the product.

Service of the product
A customer is buying a product because of that product's features and benefits along with their emotional attachment or feeling about the product. These can be slightly influenced by the GM.
The General Manager can influence the relation ship with the customer by visiting to the Customer end frequently and keeping the timely delivery and asking his subordinates to introduce new methodology to his product to improve his quality.

Customer requirements This is the first important step of the sales. Listening and asking intelligent and interesting questions that are open-ended, gather information, and discuss or communicate to his subordinates.

Reliability of the product With the help of Sales and marketing engineers monitors the reliability of the product in the market. What is the trend now and how long it will stable. Who are they competitors for us? How they are marketing and the price competition with them are important.

Marketing planning
The marketing will also be a Planning process. The long term and short term forecasting will be required to plan. In

So at least 10% of the machining time, we have to add as idle time

Total machining time = 3.24 min
Chamfering time   = 10 seconds
So Cutting time = 3.34 min.

See for example CNC lathe cost per hour = 100 Rupees

Machining cost = 100 / 60 X 3.34 = 5.56 Rupees / Piece

10 % of idle time = 0.334 min
Idle hour cost   = 100 / 60 X 0.334 = 55 Paisa
If you are producing 1000 nos per day, Cost = 55 x 1000 = 550 Rupees/day

All the above factors put together an Estimation engineer can quote to the customer. However it may be, the idle time considering or not is up to your choice.

Another idle time that are, no operator, machine break down, Power shut down, no tool, forklift, crane, and inspection are not considered. Those are all the indirect cost, which will effect the total turn over of the company. Management will focus more in this issue to increase turn over to avoid purchasing of machines and increase their Burdon.

Slot milling (Key way milling machine) with HSS.Cutter
A cutter diameter is 10 mm and no of teeth says 2 nos

Sales Planning

Slot depth is 5 mm and length is 50 mm

$$V = \frac{\pi D N}{1000}$$

$30 = 3.14 \times 10 \times N/1000$

$N = 30000/3.14 \times 10$

$= 955$ rpm

Table feed $F = f ZN$

$= 0.05 \times 2 \times 955$

$= 95$ mm / min

Cutting time $T = L / F$

$= 50 / 95 = 0.52$ min

With in a single pass it is not possible we require at least five passes

So machining time $= 0.52 \times 5 = 2.6$ min

Drilling (Bench drilling machine) with HSS. Drill
A Drill diameter is 12 mm and no of teeth says 2 nos
Drill depth is 50 mm and feed rate is 0.2 mm

$$V = \frac{\pi D N}{1000}$$

$30 = 3.14 \times 12 \times N/1000$

$N = 30000/3.14 \times 12$

$= 796$ rpm

The General Manager or the President will be the head of the total organization. The following are the important functions of the General Manager.

- Sales Planning
- Marketing planning
- Production Planning
- Customer complaints
- System implementation
- Turn over

The General manager is directly reported to the Managing Director or the Chairmen of the company.

- ➤ Increase in speed
- ➤ Reducing the total weight of the Product
- ➤ Changing the Aesthetic look of the product
- ➤ Small change in concept
- ➤ Make easy handling
- ➤ Increase in life of the product

Turn over

The Turn over is not simply a Qty or total money produced. Simply for example our turn over is one crore and the profit is one lakes excluding the expenses incurred by the company is no use. The motivation given by the management also in profit basis will give high growth to the company. The Turn over with high profit can be achieved by the above methods, which we discussed so far.

GENERAL MANAGER

The responsibility of the general manager is to manage both the revenue and cost elements of a company's income statement this means that general managers usually oversee most or all of the firm's marketing and sales functions as well as the day-to-day operations of the business. The general manager is also responsible for leading or coordinating the strategic planning functions of the company.

The general manager is also called as Chief executive officer (CEO), President, and Chief operating officer (COO), in some other companies.

Table feed F = f ZN
$$= 0.05 X 2 X 796$$
$$= 80 \text{ mm} / \text{min}$$

Cutting time T = L / F
$$= 50 / 80 = 0.62 \text{ min}$$

With in a single pass is possible so machining time is same. Include set up cost and Tool wear cost separately.

Nowadays more Tool manufacturers are available. For their tools they are recommending higher parameters so depending upon your tool purchase the speed and feeds will defer, you can calculate accordingly.

Before becoming as an estimation engineer he should know all the above. How it is possible.

- ➤ He will be an operator for at least 2 years
Then become a supervisor

- ➤ Become a problem solver
Have a Complete tooling knowledge

- ➤ Then a time study engineer
He is suitable for estimation

Planning

Planning is the process of setting objectives and determining how to accomplish them. It is, simply, thinking before doing of setting a goal and moving toward it.

Planning is very important subject, which will play a roll not only in the industry al so in life. What are all the things we can plan in an engineering industry as follows.

1. Building
2. Machining Process
3. Machineries
4. Machine Lay out
5. Production
6. Time study

1.Building

Plan the building according to the project, not simply by length and width. Before building construction, we need the following.

> Complete details of the product to be going to manufacturing.
> For those products what are all the machineries to be required.
> Depending upon the machineries how much power will be required now and in future.

2. Machining Process

We get all the manufacturing drawings with a copy of Purchase order from the sales. We can check the drawing number and revision number and date. We decide the machining process for all Work pieces. Estimate roughly the approximate time to complete the job. This is for only

> TS 16949 – Must for all Auto Mobile manufacturing companies.
> Total Predictive maintenance
> Total Quality management
> Six Sigma concepts.

What the system speaks is we have to record every thing to avoid the secondary mistakes is the concept. Manager will have a power, more engineers, Supervisors, assistants and management support too. Today's implementation is tomorrow's fruit for us.

Time study analysis

As a Manager, we have to analyze the time taken for the work piece by the following methods.

Synthetic time – The time, which will be calculated theoretically

Actual time – Which will be measured at the time of running. Its also called run time.

Arrival time – Based on the above two method we can arrive average time which one is the right time with considering all allowances like, tool change, idle run or tool approach, operator fatigue etc. the time taken for the work piece is available is not the matter, we re estimate again and again will gives us the profit. The estimation will defer person to person. Therefore, we have to involve in each aspects so many times.

Re engineering activities

This will contain the following activities.

> Change in Process
> Changing of material

environmental work place too. Another one is the Keizen activity, which is achieved by the following methods.

> Changing or shuffling the process
> Introducing the advanced tools
> Combining the operations
> Improving the skill of the Subordinates

Rejection analysis

This process will give enormous cost saving. We can do route cause analysis by asking and answering the questions like Why? When? In addition, how? Because of rejection, we will loose the costs of the raw material, process, Time, Delivery, and Profit.

Why? – is it no pool proofing method adopted?
When? – is it while machining clamping pad loosening?
How? – Every thing is correct. Then is it machine is the problem?
Ask not "Why" and ask "Why not"

System implementation

With out implementing any system we will not survive and the pressure will rise for every body. The following systems are available. For example our body is high systematic than ever and any one of the system fails in our body we will become sick. If we will get treatment in the initial step
It will avoid the serious illness.

> ISO 9000- 2000 system will be common for every Company.

to starting the project. The following are points are considered to write machining process.

> What are all the machines involved.
> Sequence of machining operation.
> What are all the critical parameters available is identified with star marks.
> Specific reference points given in the drawing should be noted. Some drawings they will give reference points to take as a reference for machining as a note will al so identified with star mark.
> Give standard time by finding out with calculation method.

While implementing the project, we have to establish the process with other departments like Production and tool design. We have to arrange for a trial run and see whether the Work piece is done with the same process or for using the Jig and Fixtures change in process is required. The production has changed your process or not to be followed. If the process is changed should be up dated. To avoid this problem, should be done combining with concern departments will give better results.

3. Machineries

Once we arrive the process and approximate time, we can find out the requirement of machines easily now. Collect repeated machines and its hour. For example per day, we are utilizing 20 hours.

For 25 days 500hrs (per month)

The following are the machineries we have selected.
Vertical turret lathe VTL – 1200 hours.
Horizontal boring machine HB. – 800 hours
Milling machine – 450 hours
Drilling machine – 900 hours
Vertical machining center (VMC) – 1200hours
Now we can buy two VTL and two Horizontal boring machines.
As per our calculation 2 VTL, consume 1000 hours only.
For balance 200 hours, we do not buy one more VTL. The spare capacity will be available in HB machine can be used for time being.

In addition, we can buy one Milling machine, two Drilling machine and two VMC. We have estimated the time approximately only.

That is not a problem. Some spare capacity is available in drilling machine and milling machine. We can do pre drilling and pre milling operations to reduce the VMC machining time.

4. Machine Lay out
We have now the entire machine details in our hand. Is it is enough? In addition, we have to plan other process as if spaces for In coming material storage, Inspection, Stores, Tool crib, Hydro testing, Scrap area, Debarring, Painting, Welding area, Final inspection and dispatch etc.
We consider all those details and plan the flow of machineries. What is flow of machineries? This will decide the Work piece movement directly and save enormous cost

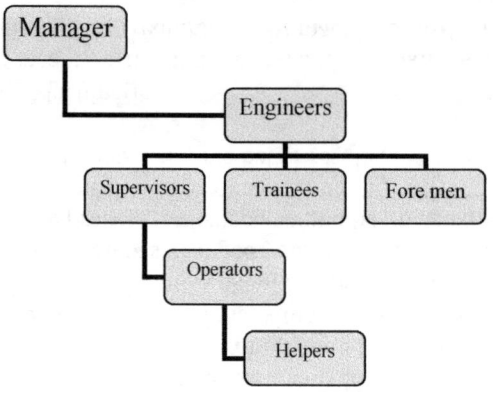

The changing of operations will change the hour rate will give profit. We do not just like that change the operations and we have to compare the final cost of the Work piece with the previous one.

Production Planning
We can plan the production neither by the Customer basis nor by the turn over basis. In turn over basis, allowing instant customers under no load condition is ok. Preference might be given to the Customer basis is good.

Cost reduction analysis
Implementation of 5'S will give direct or indirect cost saving and it is a good concept will provide good

- Production Planning
- Cost reduction analysis
- Rejection analysis
- System implementation
- Time study analysis
- Re engineering activities
- Turn over

Process planning

Already process will be established for every products or Work pieces can be re structured. We can use our experience to re establish the process will reduce the cost of the product. How we re structure the process with cost effectiveness? We can do by the following method

- Introducing Proof machining before going to CNC
- Instead of milling change in to Turning
- Instead of Turning to 09 operation
- Eliminating one operation

due to internal transport. For example in a valve manufacturing company, the following is the cumulative process we got from our total product.

First operation HB proofing 3 flanges.
Second operations do proof machining three Flanges in a VTL.
Third operations Drill tap at radial drilling
Fourth operation Hydro testing
Fifth operation Finish machining at VTL.

Now we will arrange the machineries in the following sequence. HB, VTL, RD, Hydro testing, same second machines opposite to it, and balance machines are next to the opposite bay. This will take less time to take transport. Opposite of opposite will not make more time; it will take same time to transport to next machine.

We have to plan the painting booth in a corner or provide it separately is better. This will collapse the working environment. Space for Electrical items, Panel boxes, Movement of the crane and its specifications, trenches, toilet, bathrooms and space around the Company are provided with the respective Electrical, Mechanical and Civil engineers consultation.

5. Production planning

Production planning is nothing but a monthly turn over of the company should be planned. The following are the points required for monthly planning.

- PO. From sales with Drawings.
- Process flow diagram.

- Exact time required for manufacturing the items.
- Availability of total Man-hours.
- Availability of total machine hours.
- Availability of materials.

The above 6 points are the source availability in side the company. There is no enough capacity is available, we have to out source the products thro' out sourcing department is the easier way and think about the expansion of the company by the repeatability and the stability of the orders in future.

6. Time study

We have done approximate estimation while purchasing the machineries and done the exact time calculations. Then why we can make time study repeatedly?

Time study is the separate department that is handling by the planning department or we can severally allocate a planning person for time study. Time will play the main roll not only in the company it is in every body's life also. The time study responsibilities will as follows.

- The first method is Theoretical time calculation. Calculation method is available with Estimation engineer chapter.
- The second method is to verify the time with stopwatch in machines whether they are achieving or not. Based on theory, the exact non-cutting time like tool change, tool approach and operator fatigue are not available. We have to consider those timings in second method and revise it.

The Tool crib is also maintained in the same fashion as discussed above by the Tool making or Development department.

Jig and Fixture store

The Jig and Fixture is also maintained in the same fashion as discussed above by the Tool making or Development department.

However, one major issue is the servicing and calibration of the items. We can introduce an Tool repair card for this which will contain worn out parts list like Wear pad, Clamping pad, Bushes, locating pin, Butting pin, Bolts, nuts, and a remarks column. If such parts are worn out, they will tick the part in that repair card and give it to the store. So as it is easy to repair and give back to the next issue. Major repair is written in the remarks column and intimated to tool making to avoid delay.

MANAGERIAL FUNCTIONS

The Supervisor with 7 or 8 years experienced people will become as an Engineer. The Engineer with 4 or 5 years experienced people will become as a Manager. While become as an Engineer his Burden will come down to 50% and after promoted as a Manager his burden will still gets down. We can simply say that the Manager will act as a good teacher for his subordinates.

The following are the function of the Manager.
- Process Planning

We have to maintain the Delivery Chelan and Bills properly with the computer by easy retrieval of data's.

> Issue of material

We have to see in many occasions, the peoples will come in a hurry to get the material from stores before approval of the material by QC. Some times, it will work out but not all the times and it will become much headache to the stores to closing and recording of such items.

These types of accumulated incidents will become headache at the time of year-end. There are more possibilities to forget to record the issue. Because of it, receipt and issue will not tally.

> Keeping of records

The records will consist of Delivery Chelan, Bills, PO reference, etc. will be kept separately with a separate record room.

> Implementation of 5' s concept

The 5'S concept is already given in Maintenance Engineer chapter. Please refer the same.

> Implementation of the System

The computerized system will be must for the stores. We can introduce and implement tally, or AS400 or some other system soft ware. This will help to find receipt and issue of data's date wise, party wise, Chelan wise, PO wise, etc. If the system is linked with online with planning, production, and all the departments, it is easy to find the stock of the materials and ordering of the materials.

Tool crib

> Change in process. This will give 50% of saving in time some times.
> The Time study engineer is trained in a various machines and after that, only he will become a time study engineer. Other wise it will not be possible to us to follow the above steps.

Few simple examples of the Machine and Process planning are as follows.

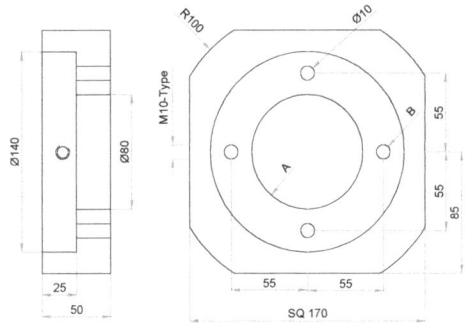

Machining process

For the above work piece, we can decide the machining process.

➢ Hold on four Jaw chuck, true the faces evenly and do face roughing, do bore roughing 75 & 130
➢ Hold on four Jaw chuck with second side, butting with chuck face and do face roughing. Finish bore to dia.78 mm and face with 2-mm allowance in thickness.
➢ Hold on four Jaw chuck with first side, butting with chuck face and do face roughing by truing the 78 bore. Finish bore to dia.140 mm, dia. 80 mm, thickness, and chamfer.
➢ Drill dia.10 – 4 nos.
➢ Drill and tap M10 – 4nos

In this above machining process, for 1, 2 and 3 fixtures are not needed for a center lathe provided with a 4-jaw chuck. However, if the quantity is in bulk, we can go for a CNC Lathe. There we require a Turning fixture.

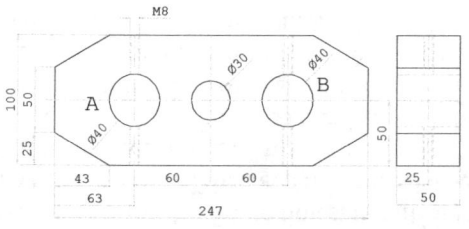

This is a casting block. So what will be the machining operations?

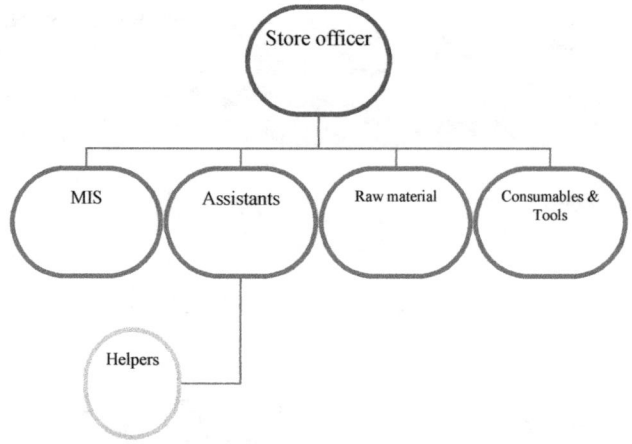

The 5's system will be implemented first and every item should be stored in the racks with containing the list. The list consist of details like, Item description, Qty, size, Rack no, Which row, Which column, will give easy to find the items.

➢ Life of the material
The some materials like resins, paints, welding rods, Coolant oils, abrasive wheels, and etc.will have certain period of lifetime only. That can be monitored with the computer system by entering the date of purchase and expiry date and putting serial numbers. Therefore, it is easy to issue-to-issue before the expiry date.
➢ Receipt of material

Tool crib in which the Cutting tools and holders are issued with taken back policy to the shop floor. Once tools where approved to be issued to the Tool crib by keeping some stock with the main stores.

The first 4 will come under Store officer control and the balance come under Tool making or Development department control.

Responsibilities of a Store officer

➢ Responsible for in-ward and out going materials
Incoming materials – the material that will come for manufacturing and assembly purpose.
Out going materials will be sent on returnable basis like, tools, measuring instruments, Hand operated machines etc.

➢ Storage of materials
The what ever may be the incoming materials are inspected and kept ready for the usage will be ready with an ease of issuing facilities.

➢ Facing one side, drill, and bore to 28 mm and chamfer in a lathe.

➢ Second side facing to finish thickness 50 mm and finish bore 30mm and chamfer in the same lathe.

➢ Drill dia. 40 – 2nos. in a Radial drilling machine by a plate Jig locating center.

➢ Drill tap M8 – 2nos. thro' – Drill jig required to perform in a drilling machine.

This is an economical process.

Responsibilities of an Operator

In the Introduction, it self I told that, this Book is written for Engineering Students, Engineers, and Employers. So first, we have to define the Functions of an Operator their field wise. In an Engineering field each and every functional areas are important.

The growth is depends all the peoples working in the company. It could be teamwork. Faster growth with in a short period is possible not because of a pure luck. It is because of the Teamwork. The following are the General machines available all over the World.

1. Cutting machines.

Cutting machines like Hand cutting, Hack saw cutting and Band saw cutting are in normal use.

2. Drilling machines.

Drilling machines like Hand drilling, Bench drilling, Column Drilling, Radial Drilling and Gang drilling machines are in normal use.

3. Turning machines

Turning machines like Center lathe, Turret lathe, Vertical lathe, and Copy turning lathe are in normal use.

4. Milling machines

Milling machines like Vertical milling, Horizontal milling, Plano milling, and Gang milling machines are in normal use.

5. Boring machines

Boring machines like Jig boring, Horizontal Boring and Vertical Boring machines are in normal use.

6. CNC. Machines.

CNC machines like VMC, HMC and VTL are in normal use.

7. Grinding machines.

Surface grinding, Bore grinding, Cylindrical grinding, Belt grinding, Vertical grinding, Profile grinding machines are in normal use.

8. Debarring machines.

Debarring machines like Hand grinder, Belt grinder and Bench grinder are in normal use.

The following Flow diagram shows the Responsibility of an Operator. In addition, we can discuss machine wise.

The Flow diagram shows the Operator first should knows to study the Drawings. We can ask, is it possible to work with out drawings? Yes, it is possible. We can ask an

companies at the year-end. It will never tally with the incoming material to out going issue. We can recruit an MBA Graduate at this place is a right person. The Store consists of the following.

➢ MIS – Material inward store

Where the material first will come to check the quality requirements and if it will found ok then pass on to concern departments via quality and store approval.

➢ Raw material store

The raw materials like castings, Steel rods, Pipes, L-angles, Mild Steel sheets etc. will come directly to the stores. Castings are sample inspected and other materials they will cut end bits and send for checking the specifications to the labs and if accepted will be stored in side or return to the party.

➢ Consumable goods store

The consumable goods like Hardware items, notes, books, and pencil etc.will are available with this store. The hardware items will be stored with Quality control approval only.

➢ Tools Store

The Tooling items like, Tool holders, Inserts, Cutters, Drills, Reamers, etc. will be stored after QC approval.

➢ Jig and Fixture store

For keeping the JIG and Fixtures more space will be required. More over it will be issued every day and taken back for servicing. It is not possible to keep it in the main stores. After the QC approval issued to Jig and Fixture store.

➢ Tool crib

- ➢ Time spent on retrieval of information
- ➢ Non availability of correct on line stock status
- ➢ Customer complaints due to logistics
- ➢ Expenses on emergency dispatches and purchases

Pillar 8 – Safety and Health Environment
Target:
- ➢ Zero accident,
- ➢ Zero health damage
- ➢ Zero fires.

The Major aim is to create a safe workplace and a surrounding area that is not damaged by our process or procedures. This pillar will play an active role in each of the other pillars on a regular basis.

A committee is constituted for this pillar which comprises representative of officers as well as workers. The General Manager Technical heads the committee. Utmost importance to Safety is given in the plant. Manager Safety is looking after functions related to safety. To create awareness among employees various competitions like safety slogans, Quiz, Drama, Posters, etc. related to safety can be organized at regular intervals.

Note- the **TPM** concept is not discussed in detail. Separate books are available to know in detail and please go through it.

STORE OFFICER

The store maintenance will be the very big headache everywhere. The stock closing function will be there in all

Operator by giving a Shaft or Square piece to turn to diameter 25 mm or Mill the Square piece to 25 mm. It will work some times, but not many times. If he will finish by 24 mm. What we will do? Nothing can be done.

The Operator should take the responsibility to do work with drawings. Through knowledge of drawing study is most essential. Missing dimensions to be noticed and informed to the supervisor do not do with assumptions. UN availability of him stops production.

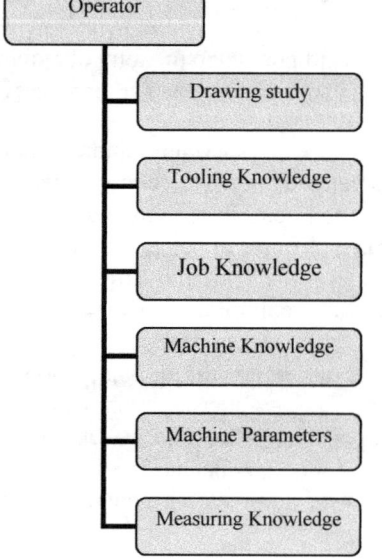

In a Xerox, copy of a drawing 26 mm shows 20 mm, because of darkness or more dirt shown. If it is in Doubt, check the Original drawing.

Tooling Knowledge

In general, Operator should know what cutting tools to be required to his machine.

operators to upgrade their skill. The goal is to create a factory full of experts.

Policy:

- ➢ Training for improvement of knowledge, skill, and techniques.
- ➢ Creating a training environment for self learning based on felt needs.
- ➢ Training curriculum / tools /assessment etc
- ➢ Training to remove employee fatigue and make work enjoyable.

Target:

- ➢ Achieve and sustain downtime due to want men at zero on critical machines.
- ➢ Train the employees for upgrading the operation and maintenance skills.

Pillar 7 – Office TPM

Office TPM should be started after achieving the first 5 pillars. Office TPM is to be followed to improve productivity, efficiency in the administrative functions and it includes analyzing processes and procedures towards increased office automation. Office TPM addresses twelve major losses to be reduced or corrected. They are

- ➢ Processing loss
- ➢ Cost loss
- ➢ Communication loss
- ➢ Idle loss
- ➢ Set-up loss
- ➢ Accuracy loss
- ➢ Office equipment breakdown
- ➢ Communication channel breakdown, telephone and fax lines

- ➤ Preventive Maintenance
- ➤ Breakdown Maintenance

Preventive and Break down maintenance already we have discussed.

- ➤ Corrective Maintenance

   The reengineering of equipment with design will improve reliability so that the preventive maintenance can be carried out reliably.
- ➤ Maintenance Prevention

Achieve and sustain availability of machines with Zero equipment failure and break down. Reduce maintenance cost by 20 %

Pillar 5 – Quality Maintenance:

Its aim is to satisfy the customer delight through highest quality with defect free manufacturing. Focus is on eliminating non-conformances in a systematic manner, much like Focused Improvement.

Policy:
- ➤ Defect free equipments.
- ➤ Prevent defects at source
- ➤ Implement poke-yoke. (Fool proof system)
- ➤ Total operator quality assurance.

Target:
- ➤ Achieve and sustain customer complaints at zero
- ➤ Reduce in-process defects by 50 %
- ➤ Reduce cost of quality by 50 %.

Pillar 6 - Training

It is aim is to make a multi-skilled employee who has to come eager to work and perform all required functions effectively and independently. Education is given to

Job Knowledge

How to do? What are the settings will be required. Is it possible to do in his machine?

Machine Knowledge

Machine knowledge is nothing but to handle the machine fluently.

Machine parameters

Length, Width, Diameter and Height to be in finger tips. Do not say after cleaning the machine weighting for the job 30 minutes and say it will not accommodate in my machine.

Measuring Knowledge

What ever may be the Job produced in the machine. The first responsibility goes for the Operator. He has to measure perfectly and Cross check with the Quality control.

1. Cutting machine

Cutting machines like Hand cutting, Hack saw cutting and Band saw cutting machines are normally available. Cutting machine any Layman can operate. Is it correct for an Engineering company? No, it is not correct. How? We can discuss.

Drawing Knowledge

The drawing knowledge is not required for this operator. Most of the items like Shafts, L- Angles, Beams, and Flats will come for cutting. The Planning department will give the computerized format by stating the material and its length, width, diameter and qty of the items.

Tooling Knowledge
He should know the life of a Bade. He should count how many jobs cutting per blade. To replace the blade before damage .Other wise the teeth will worn out and regrinding will not be possible.

Job Knowledge
Simply say the elimination of wastes. In today's assignment 25 mm road to be cut in different lengths. So calculate how many pieces to be cut with the minimum wastage of end bit by combining the lengths and quantity's.

Machine Knowledge
Machine knowledge is nothing but to handle the machine fluently. In addition, daily maintenance likes lubrication and cleaning to be done.

Machine parameters
Cutting parameters like minimum and maximum diameter, width, to be cut to be known.

Measuring Knowledge
Should knows to handle scale, tap and Vernier caliper is essential.

Cost cutting analysis
In a cutting machine what will be the profit we will get. For example today in a cutting machine, we are cutting 25, 30,

➢ SHITSUKE – (Self discipline)
We educate the self-discipline among the employees of the organization. This includes wearing badges, following work procedures, punctuality, dedication to the organization etc.

PILLAR 2 - JISHU HOZEN (Autonomous maintenance)
This pillar is geared towards developing operators to be able to take care of small maintenance tasks, thus freeing up the skilled maintenance people to spend time on more value added activity and technical repairs.

PILLAR 3 - KAIZEN :
("Kai" means change, and "Zen" means good) It will be carried out on a continual basis and involve all people in the organization. Kaizen requires a little investment.
Kaizen Policy
➢ Zero losses in every activity.
➢ Cost reduction targets in all resources
➢ Focus of easy handling of operators.

Kaizen Target
Achieve and sustain zero loses with respect to minor stops, defects and unavoidable downtimes with 30% manufacturing cost reduction.

Pillar 4 – Planned Maintenance
We can have trouble free machines and equipments producing defect free products for total customer satisfaction. This breaks maintenance down into 4 groups which are as follows.

Predictive maintenance

Total Productive Maintenance (TPM) is a maintenance program, which involves a newly defined concept for maintaining plants and equipment.

Down time for maintenance is scheduled as a part of the manufacturing. In this method, the service life of important part is predicted based on inspection or diagnosis, in order to use the parts to the limit of their service life.

TPM is an innovative Japanese concept.

The Total Predictive maintenance will consist of 8 pillars are as follows.

PILLAR 1 - 5S:

TPM starts with 5S. Problems cannot be clearly seen when the work place is unorganized. Cleaning and organizing the workplace helps the team to uncover problems. Making problems visible is the first step of improvement. The 5S concept is as follows.

➤ SEIRI – (Sort out)

Sort out the items as critical, important, frequently used items, useless, or items that are not need as of now. Unwanted items can be salvaged.

➤ SEITON – (Organize)

The Storage place is decided for every item in which other items will not be placed. The items should be placed back after usage at the same place.

➤ SEISO – (Cleanliness of the workplace)

This involves cleaning the work place free of burrs, grease, oil, waste, scrap etc.

SEIKETSU – Standardization of work place, Machines, Gangways neat and clean. These standards are implemented for whole organization.

32, and 40...100 mm rods at an average of 100 nos. The planning man studied the drawing carefully and given 2 mm per side as a machining allowance.

Some times the machining people will come and ask that the machining allowance will not be sufficient. So the additional 2 mm where added per side by the cutting machine operator un officially including the Planning staff what will happens. We can take an average of 40 mm diameter to 100 mm length with qty. as 100 nos. as an example.

Weight of the rod for dia.40 X100 = 98 grams
Cutting with excess length of 2 mm per side. Then the weight of the rod is = 102 grams
Loss of material = 102 – 98 = 4 grams
Loss of material per day = 4 x 100 = 400 grams
For one month = 400 X 26 = 10.4 kegs.
Material cost   = 30 x 10.4 = 312 Rupees.
Machining excess material per side 30 seconds
Machining cost = 1min. X100 nos.X26 days
                    = 43.5 hrs.X 60 Rupees. = 2610 Rupees.

Total Loss per month = 2610 + 312 = 2922 Rupees.
This is for single cutting machine and one item. Then we can calculate for no. of machines and quantity it will come enormously.

Route cause analysis

We found a problem that, the machining allowance required is more. To avid this we can check the cutting taper. The cutting taper will comes because of the in sufficient clamping pressure and seating surface of the job will be corrected.

In manufacturing of castings, also they are providing more machining allowance. They feel that more weight will give more Turn over. That is not correct. We are producing 100 tons will become 105 tons. However, after a week time if the casting price will increase, we will loss 35 tons increase in price.

In addition, every body losing the materials in to scrap, the Raw material will become demand. This is the loss for the Industry and the Country.

2. Drilling machine

Drilling machines like Hand drilling, Bench drilling, Column Drilling, Radial Drilling, and Gang drilling machines are available.

Drilling is easily the most common machining process. Maximum of all metal-cutting material removed comes from drilling operations. Drilling involves the creation of holes that are right circular cylinders.

The chips must exit out of the hole created by the cutting. Chip exit can cause problems when chips are large and/or continuous. For deep holes in large work pieces, coolant may need to be delivered through the drill shaft to the cutting front.

Drawing Knowledge

He knows to study the drawing and see what are the holes are to be drilled and reamed.

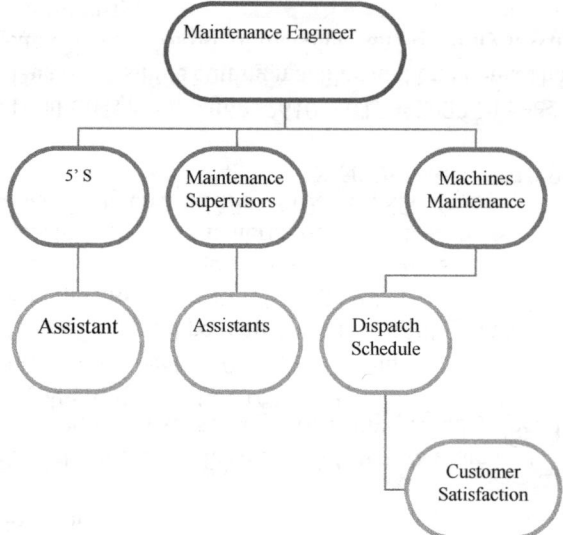

Periodic maintenance

This will be done periodically at certain interval of time depends on the life fixed. This called a time-based maintenance consists of

➢ Periodically inspecting the equipment
➢ Changing the Coolant Oil
➢ Changing the Lubricating oil
➢ Servicing and cleaning equipment and replacing parts to prevent sudden failure and process problems.

out in the machines like continuous production line. We can do such a maintenance in cases like hand grinder, bench grinder, etc. Which we can repair or buy easily.

Preventive maintenance
This is a daily maintenance like,
Cleaning and Oiling – Every day opening of the shift and the closing of the shift we have to Clean the machine bed surface and all the guide ways by waste and put oil for smoothness. While doing this, the machined metal powders sticking on the slides will remove and the rust will not form. It will improve the machine life too. We have to put grease every day through the grease nipples wherever it has to be provided.

Tooling Knowledge
Should know about the usage of drills, Reamers, C' sink cutters, Hole mills and C' bore cutters. In addition, its speed feed selection and usage of Jigs.

Job Knowledge
In which side it is coming and is there any orientation is required. Is it is suitable for his machine.

Machine Knowledge
What are the feed, speed, Run out of the spindle available in his machine and accuracy of the machine.

Machine parameters
Parameters like minimum and maximum diameter of the drill size to be loaded.
What are the length, width, and height of the job to be loaded in his machine.
What is the maximum weight of the work piece to be loaded in his machine.

Measuring Knowledge
Should knows to handle Vernier caliper, Micrometer, Depth gauge is essential.

Cost cutting analysis
- ➢ Reduce no. of Drills per drill hole
- ➢ Use Proper Tooling.
- ➢ Use Jigs wherever required.
- ➢ Use Quick changers.

Route cause analysis

It is required where ever the positional accuracy and Pinholes are coming. To get Positional accuracy and repeatability avoid marking and use Jig. Pin holes or Orientation holes are made by the sequence of drilling, then use Hole mill and Ream. Adding a hole mill is not a loss. It will save the Job and will give more accuracy.

3. Turning machines

Turning machines like Center lathe, Turret lathe, Vertical lathe, and Copy turning lathe are available.

Turning is another one of the basic machining processes. Turning produces solids of revolution, which can be tightly tolerances because of the specialized nature of the operation.

Turning is performed on a machine called a lathe in which the tool is stationary and the part is rotated. The term "facing" is used to describe removal of material from the flat end of a cylindrical part. Facing is often used to improve the finish of surfaces that have been parted.

Drawing Knowledge

He knows to study the drawing and he under stands that, about his operations like Facing, Turning, Boring and its surface roughness requirements.

Tooling Knowledge

Cutting parameters of a single point tools and Boring tools

Job Knowledge

Loading and un loading of the Job and its setting.

available for each method. We learn more depends upon the system followed in your company. We do not get confused now by putting together all the systems.

MAINTENANCE ENGINEER

The Maintenance department is the Key of the Company. Before going to the maintenance, we can also discuss one fact that, In Western Countries they are procuring advanced new machines and fixing the Life of the machine and working itself. In India, most of the companies are buying the older once that are out dated and Life expired is not a correct way of approach to implement the system. It is difficult to implement the predictive maintenance. The Maintenance Engineer is responsible for the following activities.

- ➢ Break down maintenance
- ➢ Preventive maintenance
- ➢ Periodic maintenance
- ➢ Total Predictive maintenance (**TPM) is an innovative Japanese concept. This** will have 8 pillars of its activities and can be achieved by the teamwork only.

Break down maintenance this kind of maintenance will be done only if the machine will become failure. After getting the intimation from the shop floor by orally or in written format we will attend the problem and solve it. This could be used when the equipment failure does not significantly affect the operation or production or generate any significant loss other than repair cost. This will not work

The Quality Circle is a small group of employees from the same work area who voluntarily meet at regular intervals to identify, analyze, and resolve work related problems. This can not only improve the performance of any organization, but also motivate and enrich the work life of employees. The use of Quality Circles in many highly innovative companies has been proven.

Six Sigma

Six Sigma is a process improvement methodology originally developed by Motorola to systematically improve processes by eliminating defects. Defects are defined as units that are not members of the intended population. Since it was originally developed, Six Sigma has become an element of many Total Quality Management (TQM) initiatives.

Total Quality Management is a management strategy aimed at embedding awareness of quality in all organizational processes. Total Quality provides an umbrella under which everyone in the organization can strive and create customer satisfaction.

Total Quality is a people focused management system that aims at continual increase in customer satisfaction at continually lower real costs.

Quality Assurance through statistical methods is a key component in a manufacturing organization, where TQM generally starts by sampling a random selection of the product.

For Quality Circle, Six Sigma, TS 16949 and TQM methods we can go elaborately is not possible now. The separate books are

Machine Knowledge

What are the feed, speed, Run out of the spindle available in his machine and accuracy of the machine.
Mainly he should know the thread cutting operation.

Machine parameters

He should know the Diameter of the Chuck, Center Height, and maximum distance in between the centers.

Measuring Knowledge

Should knows to handle Vernier caliper, Micrometer, Depth gauge, Bore dial is essential.

Cost cutting analysis

> Reduce the setting time
> Use Proper Tooling.
> Use Turning fixtures wherever required.

Route cause analysis

It is required where the run out and concentricity is coming.
It could be achieved by adding one proof machining operation. In addition, introduce a Turning fixture. In case of a Poor finish, change speed, feed and Insert and give a finish cut with less allowance at higher speed. This will be an addition of one more operation but it will save the work piece and saves more cost.

4. Milling machines

Milling machines like Vertical milling, Horizontal milling, Plano milling, and Gang milling machines are available

Milling is as fundamental as drilling among powered metal cutting processes. Milling is versatile for a basic machining process, but because the milling set up has so many degrees of freedom, milling is usually less accurate than turning or grinding unless especially rigid fixturing is implemented.

For manual machining, milling is essential to fabricate any object that is not axially symmetric. There is a wide range of different milling machines, ranging from manual light-duty to Heavy duty. Manual mills are common in job shops that specialize in parts that are low volume and quickly fabricated. Such job shops are often termed "model shops" because of the prototyping nature of the work.

The parts of the manual mill are, the knee moves up and down the column on guide ways in the column. The table can move in x and y on the knee, and the milling head can move up and down.

Drawing Knowledge
Study the drawing and see what are the operations involved in Milling. The important geometrical requirements like flatness, surface finish, Parallelism, and angularity are noted carefully.

Tooling Knowledge
He will knows the Cutting parameters of a Multi point tools like End mill, Slot mill and Face mill cutters.

Rejection analysis
The meeting for Rejection analysis conducted in shop floor with concern department persons with all department heads. The rejected materials are kept in the shop floor to its allocated area. We visit one by one with all heads and analyze the problems as below
> Raw material specification and hardness
> Process capability
> Route cause analysis and its implementation

This will give cost effectiveness to the product.

Quality circle
A quality circle is a volunteer group composed of workers who meet together to discuss workplace improvement, and make presentations to management with their ideas. Typical topics are improving safety, improving product design, and improvement in manufacturing process. Quality circles have the advantage of continuity; the circle remains intact from project to project.

Quality Circles were started in Japan 1962 (Kaoru Ishikawa has been credited for creating Quality Circles) as another method of improving quality. The Japanese Union of Scientists and Engineers (JUSE) coordinated the movement in Japan. Prof. Ishikawa, who believed in tapping the creative potential of workers, innovated the Quality Circle movement to give Japanese industry that extra edge in creativity.

documents, were introduced during manufacturing, planning and control.

Management had to confirm all operators are equal to the work imposed on them on holidays, celebrations and disputes did not affect any of the quality levels.

Inspections and tests were carried out, and all components and materials, bought in or otherwise, conformed to the specifications, and the measuring equipment was accurate, this is the responsibility of the QA and QC department.

Any complaints received from the customers were satisfactorily dealt with and the corrections carried out are informed to the customer immediately.

Feedback from the user or the customer is used to review designs. If the original specification does not reflect the correct quality requirements, quality cannot be inspected or manufactured into the product.

Customer complaints What for all we doing is for to satisfy the customer. A customer complaint received by sales is handed over to QA. The QA head will call for the meeting and submit all related documents that we are going to analyze. We will analyze the following.

- ➢ Raw material specification
- ➢ Hardness of the material
- ➢ Structure of the material
- ➢ Geometrical tolerance of finished product
- ➢ Fitment of the product
- ➢ Performance of the product

If some failure in product will result any one of the above require correction. The correction will be carried out and recorded after getting the approval from the Customer end.

Job Knowledge
Loading and un loading of the Job and its setting.

Machine Knowledge
What are the feed, speed, Run out of the spindle available in his machine and accuracy of the machine. Handling of X, Y and Z travels.

Machine parameters
He should know the X, Y, and Z travel of the machine and Distance from Spindle face to Bed face.

Measuring Knowledge
Should knows to handle Vernier caliper, Micrometer, Depth gauge, is essential.

Cost cutting analysis
- ➢ Reduce the setting time
- ➢ Use Proper Tooling.
- ➢ Use Milling fixtures wherever required.

Route cause analysis
It is required where the Flatness and parallelism is coming. It could be achieved by proper clamping and arresting the rotation of the Work piece. Provide rigid clamping or a Milling Fixture. Poor finish will come because of the worn out inserts or speed and feed is not matching. Give a finish cut with 0.2 mm allowance at higher speed and lower the feed rate.

5. Boring machines

Boring machines like Jig boring, Horizontal Boring and Vertical Boring machines are available.

Jig boring machines are more precise and rigid in construction.

Precise parts of the machines are manufactured in this machine. This is called as a mother machine. In Horizontal and Vertical Boring machines, we have four axes. Those are X, Y, Z, and W axis. The 'W' axis is the rotary moment of the Table. In addition to milling, we can do precise boring operation.

Drawing Knowledge

Study the drawing and see what are the operations involved in Milling and boring.

Tooling Knowledge

Cutting parameters of a Drills, Hole mill, Reamers, Boring tools, multi point tools like End mill, Slot mill and Face mill cutters.

Job Knowledge

Loading and un loading of the Job and its setting.

Machine Knowledge

What are the feed, speed, Run out of the spindle available in his machine and accuracy of the machine. Handling of X, Y, Z, and W travel.

Machine parameters

- ➢ Fitment of the mating parts
- ➢ Run out of the shaft
- ➢ Adding the lock nut
- ➢ Adding the cotter pin
- ➢ Adding the lock washer
- ➢ Giving training for new assembly personnel.

In Automobile parts manufacturing the statistical process controls in manufacturing operations usually proceed by randomly sampling and testing a fraction of the output and by using Run chart etc.

Variances of critical tolerances are continuously tracked, and manufacturing processes are corrected before bad parts can be produced.

The quality assurance defined by the International Standards contained in the ISO 9001- 2000 series and the specifications for quality systems. Still, in the system of Company Quality, the work being carried out was shop floor inspection, which did not control the major quality problems. This led to quality assurance or total quality control, which has come into being recently.

Total quality control

Total Quality Control is the most necessary inspection control of all in cases where, despite statistical quality control techniques or quality improvements implemented, sales decrease.

The major problem, which leads to a decrease in sales, was that the major characteristics, ignored during the search to improve manufacture and overall business performance. Where Reliability Maintainability Safety Conformance to specifications i.e. drawings, standards and other relevant

Servicing and documentation

We have to see in engineering and manufacturing, quality control and quality engineering are involved in developing systems to ensure products, or services are designed and produced to meet or exceed customer requirements. These systems are often developed in conjunction with other business and engineering disciplines using a cross-functional approach.

Failure testing

A valuable process to perform on a whole consumer product is failure testing, the operation of a product until it fails, often under stresses such as increasing vibration and temperature.

This exposes many unanticipated weakness in a product, and the data is used to drive engineering and manufacturing process improvements. Often quite simple changes can dramatically improve product service, such as

- ➢ Small correction in the Pattern
- ➢ Changing the paint resistant.
- ➢ Tightening the belt

He should know the X, Y and Z travel of the machine and Distance from Spindle face to Bed face, size of the bed, Facing chuck diameter, Maximum boring diameter and maximum drilling capacity of the machine.

Measuring Knowledge

Should knows to handle Vernier caliper, Micrometer, Depth gauge, Bore dial is essential.

Cost cutting analysis

- ➢ Reduce the setting time
- ➢ Use Proper Tooling.
- ➢ Use Fixtures wherever required.

Route cause analysis

It is required where the Flatness, parallelism, Concentricity and ovality is coming. It could be achieved by proper clamping and arresting the rotation of the Work piece. Provide rigid clamping or a Fixture. Poor finish will come because of the worn out inserts or speed and feed is not matching. Give a finish cut with 0.2 mm allowance at higher speed and lower the feed rate.

6. CNC. Machines.

CNC machines like VMC, HMC, and VTL are available. The abbreviation CNC stands for Computer Numerical Control, and refers specifically to a computer "controller" that reads G-code instructions and drives the machine tool, a powered mechanical device typically used to fabricate metal components by the selective removal of metal. CNC does numerically directed interpolation of a cutting tool in the work envelope of a machine.

The number of machining steps that required human action has been dramatically reduced. Improvements in consistency and quality have been achieved and reduced the frequency of errors Numerical control machines or CNC machines can be defined as a form of programmable automation in which the process is controlled by numbers, letters and symbols. These numbers, letters and symbols are coded in an appropriate format to define a program of instructions for a particular workpiece or job. This code can be changed or altered when a different job is to be performed.

Drawing Knowledge

For CNC peoples drawing study knowledge is much more essential. While drawing study, see what are the previous operations are completed and what are all the work pending can be written in separate sheet. It will help to make the process, tooling list, programming and through knowledge of that drawing.

Tooling Knowledge

Cutting parameters of a Drills, Hole mill, Expanding Reamers, Micro Boring tools, multi point tools like End mill, Slot mill and Face mill cutters. Types of inserts to be used with different type of metal cutting process. For each of this machining methods there are certain parameters that have to be specified prior to work and are called cutting conditions. Although for each process there are certain

failures are isolated and the failures are corrected. The statistical distributions of important measurements are tracked. We can do life test in which the sample product is operated until a part fails. The failure is then isolated and engineers design an improvement.

The all above said records will be more help full to do change in design or further development of the product.

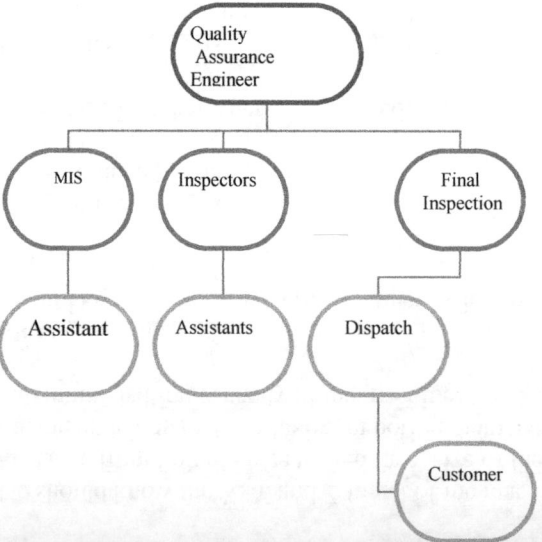

> Implementation of Total Quality management. TQM.

Production The production department where the product is machined will be monitored by the quality assurance and give solutions where they found difficulty. Say for example, the parallelism was not found ok because of the thin cross section of work piece instead of saying not ok. Rout cause for this problem is to reduce the clamping pressure etc.

Inspection To Inspect or measure the elements or a product is a regular activity of the quality control. The measuring equipment was accurate, and calibrated, this is the responsibility of the QC department.

Design and Development
The Quality assurance will help to the Design and development department by doing the following activities.

> By stating the quality of raw materials – Check the raw material specification matching to the Design specification.
> Assemblies – Check the fitment and its clearance meeting to the design requirements.
> Product – Performance of the product will be checked in co-ordination with the Design peoples.
> Components – Individual parts measurement will be recorded.
> Inspection process – Inspection method will be defined for Components, assemblies and product.
> Testing - The sample has to be tested for things that matter most to the end users. The causes of any

differences the most common parameters encountered in almost all of them are the speed, feed, and depth of cut.

Job Knowledge
Complete machining process. What are all the critical parameters and accuracy of the Work piece.

Machine Knowledge
In this machine, the feed and speed will be automatically controlled through programs. Therefore, knowledge in programming and accuracy of the machine is essential. There are basically two methods of programming or positioning:
1. Absolute programming
2. Incremental programming
   The absolute programming means that the tool locations are always defined in relation to the zero point
   The incremental programming means that the next tool location must be defined with reference to the previous tool location.
   It is must be noted though that in cases of manual part programming the absolute method should be preferred because of the fact that when incremental programming is used and an error occurs in the coding of the part program the error will be transferred throughout the program. As a result the program must be rewritten in contrast with absolute programming where only the line with the error should be corrected.

> Manual part programming
 In this program, the processing instructions are documented on a form called a part program

manuscript. This will contain the positions of the tool relative to the work piece in order to perform the required task. This listing can also include other information referring to the cutting conditions such as speeds and feeds.

➢ Manual data input

Manual data input is a procedure entered directly at the site of the processing machine. The programming is simplified to allow machine operators to do the programming instead of using part programmers.

➢ NC programming using CAD/CAM

The CAD/CAM is an advanced form of computer-assisted part programming in which an interactive graphics system equipped with NC programming software is used to facilitate the part programming task. The part programmer works on a CAD/CAM workstation and it provides the machining commands. The programmer has visual feedback on a graphics screen. In addition some programming cycles are automated to reduce the total programming time.

Machine parameters

The variety of operations were performed on the same machine, a variety of tooling was needed for this operations. The machine center possesses the feature of automatic tool change. The tools were kept on a magazine and when a specific tool was needed the machine exchanged the tool. This function is automatic and takes

are only the in puts only. The complete responsibilities of the output are come under Development engineer or an Industrial engineer.

In some of the Company's the Industrial engineer's work may differ. They expect the knowledge of machine procurement and capital goods only. Such type of persons has to enjoy the purchase of goods with good commissions.

QUALITY CONTROL ENGINEER

The function of the Quality control department is to inspect the Work pieces after completion of the process and this is a post mortem process. This will not ensure the Quality and the chances for approving bad castings also. After the ISO 9000-2000, procedure implementation it will be changed enormously everywhere in the World. Now we call it as Quality assurance department.

The Quality Assurance Engineer is responsible for the following activities.

➢ Production
➢ Inspection
➢ Design and Development
➢ Servicing and documentation
➢ Total Quality control
➢ Customer complaints
➢ Rejection analysis
➢ Implementation of Quality Circle
➢ Implementation of Six Sigma

CMM is not available with us, we can make templates and necessary gauges at the trial stage itself. The target date will be fixed for buying and making measuring instruments and gauges in the meeting.

Method of inspection

What are all the parameters to be measured in which method to be decided? If the CMM is available with us there is, no need of gauges required at the initial stage other wise we discuss the following.

Is it the Flatness and parallelism to be checked by Trimos or dial with height gauge?

Is it the Co- ordinates to be checked with Trimos or poisoning gauge?

Is it templates are required to measure profiles of the casting and machining?

Target Date The final target date to be arrived by averaging all the activities so far planned or arrive all the activities with in the target date given by the customer.

The development engineer will have assistants like Designers, draughts men, tool follow up peoples under him. He has to extract and follow the work given to them with in a target date.

The proving of the Work piece or the product will come under him. All the department staffs will help him and they

place when the program asks for a tool change. He should know the X, Y and Z travel of the machine and Distance from Spindle face to Bed face, size of the bed or chuck diameter, Maximum load capacity of the machine is the much more important point. We can use 80% of the load capacity. Machine otherwise will become damage with in a short period. We should know the origin point of the spindle and what would be the Z offset to take work offset.

Measuring Knowledge

Should knows to handle Vernier caliper, Micrometer, Depth gauge, Bore dial, Trimos is essential. A CNC operator should through in all knowledge's, just making on, off the switches will not effective, and get bored in future. No bodies prefer him and there are no more chances to switch on to other company's.

Cost cutting analysis
✓ Reduce the setting time
✓ Use Proper Tooling.
✓ Use Fixtures for clamping.
✓ Reduce the cycle time

Route cause analysis - The setting time could be reduced by automatic clamping in olden days. For that, it lacks and Lacks rupees were invested. Now every body is going for Auto matic Double pallet changing machines. Compare to Idle time its nothing.
Proper advanced tooling like Solid carbide drills, Instead of hole mills use 'U' drills, Gun drills for deep drill applications and Insert type through coolant drills.

In CNC, tooling application will play a major role. Up date the advanced tools available in the market. Just like that, do not introduce and buy tools. Make trials and compare the life. For example, one insert will cost 300 rupees. For the same application, another insert with 450 rupees is available. We can buy one sample and see the life of the insert. It will give 50% improvement then no use.

50% improvement than 300 rupees insert will give 450 rupees.

70% improvement means we will get profit as 60 rupees per insert. We consume 100 inserts per week means; the profit per month will become 24,000 rupees. So upgrading the tools and study of daily market is important.

Usage of fixtures in a mass production industry will give high profit. The work piece is a one-type non-standard job then use standard fixture element.
Reducing the total cycle time in minutes will give high profit. For example, we are using more number of drills in a work piece instead, we can cut two drills and introduce a 'U' drill will give more profit. If the core hole is available, straight we can use 'U' drill and save the cost.

7. Grinding machines.
Surface grinding, Bore grinding, Cylindrical grinding, Belt grinding, Vertical grinding, Profile grinding machines are available. Grinding is a finishing process used to improve surface finish, abrade hard materials, and tighten the

The loading and un loading will play a main roll in the machining. The machining time is three minutes and loading and un loading will be 5 minutes will also take place will be avoided before Designing a Tool.
➤ If the work piece will have a thin cross section, clamping presser and torque wrench requirement.

➤ Decide the non standard Gauges, Templates design, and its tolerance.
What are the instruments available with us? Is it enough to check dimensions and critical parameters? To reduce the inspection time we do gauges, when going for a mass production. In the initial stage, we can go for gauges, if it is mandatory only.

➤ Decide the target date for completion of the proto type. We fix the design, Fixture completion, machine hour available from planning or make free the machine at the time of trial, Production, and dispatch.

Gauges requirement- We have to study, how the critical parameters to be measured with the team members. Quality engineer will raise the questions or he has to say how he is going to measure the parameters. In initial stage, we avoid Plain plug gauges. The thread plug gauge and ring gauges we have to order is the must.

If we will have a CMM with us; there is no worry about the templates and gauges to critical parameters. We have to develop those gauges in mass production stage. If the if required, and again refresh the process capability with outsourcing.

Jig and Fixture requirements The Jig and the Fixture requirement are decided in the meeting after finalizing the machining process. The following points to be discussed to design a Jig and fixture in the meeting

> To which operation fixture is required.
In the sequence of machining operations, for the entire stages fixture will not be required. In the operation sequence, to which operation Fixture will be required to be finalized.

> Location point to the Drill jig will be decided.
For the Drill Jig any pre drilling is available in the process or is it the side reference is enough to full fill the drawing requirement and target time to design and development completion to be fixed.

> Reference points to the Fixture.
We have to see in the drawing, the customer or the designer will give reference points to perform the machining operation. We find in the drawing is good and take the same reference otherwise the reference points to be fixed by the team members. We will discuss about three-point contact and further more spring loaded points for arresting the vibration. Target time to design and development completion to be fixed.

> Loading and un loading facility.

tolerance on flat and cylindrical surfaces by removing a small amount of material
in grinding; an abrasive material rubs against the metal part and removes tiny pieces of material.
The abrasive material is typically on the surface of a wheel or belt and abrades material in a way similar to sanding.
The abrasive action of grinding generates excessive heat so that flooding of the cutting area with fluid is necessary.

Drawing Knowledge
Study the drawing and see what are the operations involved in Grinding and the type of material and its hardness.

Tooling Knowledge
Should know the type of grinding wheels and its application,
Such as Silicon carbide and Aluminum oxide wheels. The wheels are available with many combination and many colors.

Job Knowledge
How to achieve the parallelism, Concentricity, Run out and perpendicularity.

Machine Knowledge
Wheel balancing, longitudinal and cross feed adjustment and taper correction are the required skills in this machine.

Machine parameters

He should know the X, Y and Z travel of the machine and Distance from Spindle face to Bed face, size of the bed and Rpm of the machine

Measuring Knowledge
Should knows to handle Vernier caliper, Micrometer, Bore dial and dial stand is essential.

Cost cutting analysis
  ➢ Reduce the metal removal.

Route cause analysis
We are doing this process after some pre machining operation. Therefore, in the previous machining operations we can control the flatness, machining allowance and other geometries. So as the time will reduce enormously.

8. Debarring machines.
Debarring machines like Hand grinder, Belt grinder and Bench grinder are available.
Debarring is the final operation, can be done before final inspection and packing. Here there is no drawing study or machine knowledge is required. The job knowledge is most essential. The operator should know the value of the work piece and he will not damage the work piece.
Do not make any damage, scratches, uneven chamfer; uneven counter sinks on the work pieces. Most probably, these peoples will move the work pieces one place to another place, so they are educated to place in suitable places like wooden platform, Bins or specially made work holding devices.

  ➢ Target date.

Machining process We have to distribute the drawings to the team members at the date of purchase order received and give the time to study the drawing. We fix the meeting date after two days. When the meeting is started, the machining process is finalized with all peoples opinion combine together.

Critical Parameters We have to analyze the critical parameters available in the drawing like, Parallelism, perpendicularity, Concentricity, Flatness, Surface Roughness, Tolerances, and Positional accuracy.

Depends upon the critical parameters some time the machine will change that can be identified with the machining process it self. We can take more time to finalize the process is well and good. We can refine the process again and again up to the end of meeting is good.

Availability of Machine hours We have to check with the planning and production engineers, whether the machine hours is available with us. If the machine hours are not available then we can go for out sourcing.

We initially developing all the tooling for trial purpose and go for out sourcing are not good. The same type of machine may be not available with outsourcing means we will become in trouble. If the machine hour is not available, call the out sourcing engineer for meeting, also call the vendor

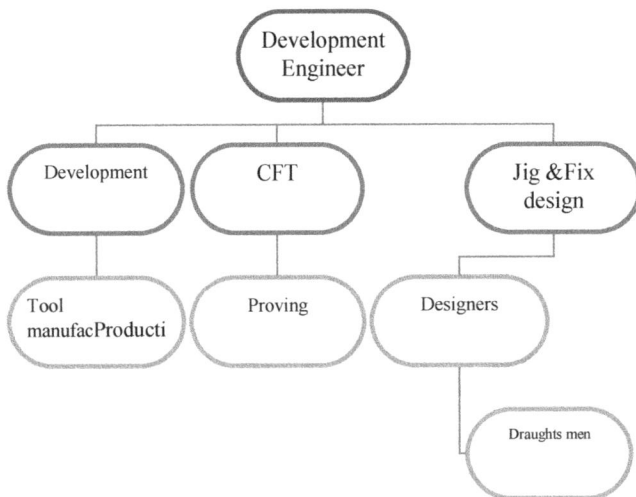

The following points will be discussed from the meeting.
- ➤ Machining process
- ➤ Critical Parameters
- ➤ Availability of machine hours
- ➤ Jig and Fixture requirements
- ➤ Gauges requirement
- ➤ Method of Inspection

SUPERVISOR Is the Star of the Company and basic qualification required is a Diploma in desired discipline.

The roll of a Supervisor is very critical and it is a basic for all other posts. A good supervisor will become a manager and a General Manager. So for all other posts this is the first step of the ladder.
In supervisory functions, first step is to know the responsibilities of an Operator. That we have already discussed in detail.

Labor Management

In supervisory functions, second step is the Labor management. No other technical skills were asked, why? We have already discussed about the labor or the operator skills.
An operator knows about Drawing study and machines thoroughly and he will come to approach a supervisor for doubts only. Therefore, supervisor can improve his knowledge himself with practical experience only. He will have subordinates like 100s of labors directly under his control. Even a General Manager will not have this much of peoples directly under him. What is the labor management? We are not discussing about the labor laws here. How to satisfy the labors with out disturbing the management will be the point.

For that we know aboutgeneral requirements of a Labor.
The Labor flow diagram shows their general requirements.
In all factories 90% of the below facilities are available.

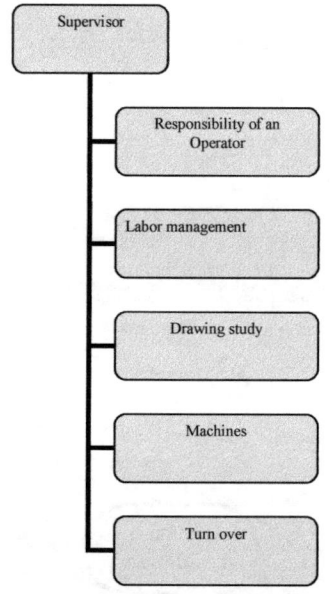

Time card – Even in Computerized factories, they
introduced as a pay cards. They will have a in – out
punching card, which is for his attendance purpose only.

The Development engineer is called Industrial engineer in
some companies. The every new product comes under his
control. He will lead the Cross functional Team (CFT)
members.

The Input for the Development Engineer comes from the
Sales Engineer with Purchase order and Drawing.
We Verify the Drawing with the purchase order like
drawing number, revision number and the target date given
by the customer.

We have to arrange a meeting with the Cross-functional
team members. Who are they team members? The teams
consist of engineers like, Planning, Production, Quality,
Development and a team leader who is going to looking
after this product.

- If we are using two bore locations, we have to use one as a round and another one as diamond locator.

- After completing the assembly drawing, tolerate the assembly requirements to avoid cumulative errors and best results.

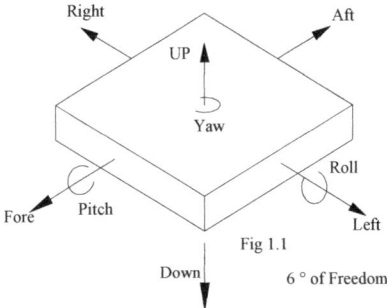

Fig 1.1

6 ° of Freedom

The above figure shows the six directions of movement or translation are called degrees of freedom in three-dimensional environments.

They are Up and Down, Left and Right, Fore and Aft, Roll Pitch and Yaw. This is an important factor for design of Jig and Fixtures. In addition, while machining, for job setting we can arrest the six directions of moments will give the correct result.

In Pay card information's like Drawing number, Description, Job loading and un loading time, Completion time and Idle time will be there. After ending of the each shift, he will fill it up and get the signature from the shift Supervisor.

Then it could be entered in to the computer and based on that he will get the salary. This will helps to avoid time fluctuation between one operator to another operator.

In small Factories Time card have information's like Ist shift, IInd, IIIrd shift and over time only will be there. The Supervisor can sign while the operator entering and while going out.

ESI. - The ESI benefits are explained briefly as under.

All non-seasonal factories using power and employing 10 or more persons and non-power using factories employing 20 or more persons are coverable under the ESI Act. All employees in such factories getting wage up to Rs.6500/-per month. are coverable under the Act. It will defer country to country.

Benefits under the Act:
The following benefits are provided to the insured persons and their family members under the provisions of ESI Act.
- Medical Benefit.
- Sickness Benefit
- Maternity Benefit
- Disablement Benefit.
- Dependant's Benefit.
- Funeral benefit.

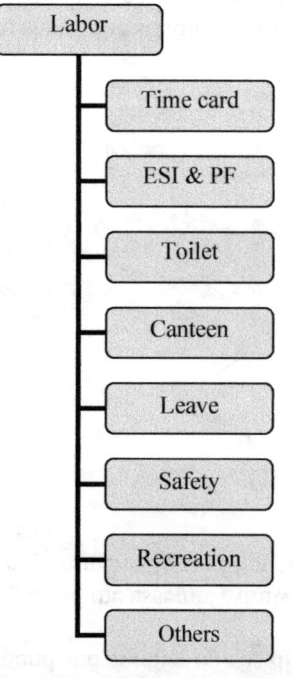

- Labor
  - Time card
  - ESI & PF
  - Toilet
  - Canteen
  - Leave
  - Safety
  - Recreation
  - Others

➢ Locator's parallelism, perpendicularity, concentricity, and cylindricity should be with in 0.01 mm allowed or you can take 1/5 times of the job tolerance as a practice

➢ After knowing the location we have to draw the component drawing and center locator.

➢ Provide clamps suitably. Then draw the outer lines, which will become a base plate.

➢ Base plate parallelism and perpendicularity should be with in 0.01 mm allowed or you can take 1/2 times of the job tolerance as a practice

➢ If locating holes are not available, give three points contact locations and give a top clamp against it. Side stoppers to be given and clamp against it for pool proofing.

➢ Spring loaded points to be added in addition to avoid vibrations while machining.

➢ Clamps to be slotted and spring loaded and stud length should be not more than 1 or 2mm after clamping the job to avoid handling time.

➢ Fix wear pads for long life and avoid more contact areas for less friction and easy handling.

Fixture. Because once the Jig and Fixture is proved, any body can do in the same method.

The following are the Guidelines for designing a Jig and Fixtures

Jig – This will locate and guide the work piece to facilitate manufacturing

Fixture - This will locate and fix the work piece, but will not guide the Work piece.

> Study the drawing and collect the machinery details in which we are going to produce work pieces and using the fixture or jig.

> Study the working area X, Y, and Z travel of the machine is suitable or not to produce the job.

> If the machine is suitable, find the location point.

> Location should be decided where the dimensions to be distributed as an origin point.

> Once location has found, fix our locator diameter 0.02 less than the minimum diameter of the bore size or h6 tolerance of the minimum diameter

> Fix the length of the locator size as one third of the bore length of work piece

The First benefit i.e. medical Benefit is provided by the State Government through the ESI Scheme and the remaining five benefits, which are cash benefits are provided by the ESI Corporation directly. Therefore, The ESI is the very good facility provided by the government and will provide more security to the entire family. In present condition minimum 500 rupees for medical expenses is required to entire family.

Provident fund (PF) - It is a mandatory, tax-qualified, defined contribution, retrial benefit plan wherein equal contribution at the rate of 12% is made by the employer and the employee and the same is payable in lump sum on retirement. The current rate of PF contribution by a member is 12% of Salary / Wages (Basic + Dearness Allowance) with matching contribution from employer. Settlement can be done only after a waiting period of two months from the date of resignation but in cases of members leaving abroad, settlements can be done immediately and settlements are immediate in case of female members who resign from the services for the purpose of getting married.

Gratuity - It is a mandatory tax-qualified, defined benefit plan paying ½ months salary for every year of service / work completed in lump sum at retirement. Every employer employing more than 10 persons must pay gratuity to his employees, on the discontinuance of service, if they have served a minimum of 5 years of continuous service (except in the case of death).

Toilet – One can ask, is there any company with out Toilet. That is not a matter. How it could be cleaned and kept. Most of the diseases starting from the toilet only. Diseases like Fever and Malaria will come easily. That is the very big indirect loss for the Company. He will take leave often, realize, and switch over to some other company.

Canteen – For example in a Company Labors are taking more leaves and leaving every month with out information. What will be the rout cause analysis? Go to Human Resources department get fast 6 months resigned peoples address and find how far away they are coming. If the distance is more, he will find difficulty to get food.
He is a bachelor and has some other nativity and taking room nearer to the company. He does not know to cook or did not find enough time to cook or cooking items are not available near by. Therefore, we provide canteen and enjoy the benefits both.

Leave – There is an earn leave and sick leave is mandatory to give all the employees. Leave is essential for every body and don't refuse every time. The essential leaves like marriage function, Illness and their relatives' death etc. The turn over is calculated with considering these leaves also or provide alternate arrangement by considering this.

Safety – We have given the appointment to an operator with PF, ESI, good Toilet, Food facilities with leave. Now he is coming to work with satisfactory.

However, he met an accident inside the factory because of a safe guard not provided in his machine. This is a

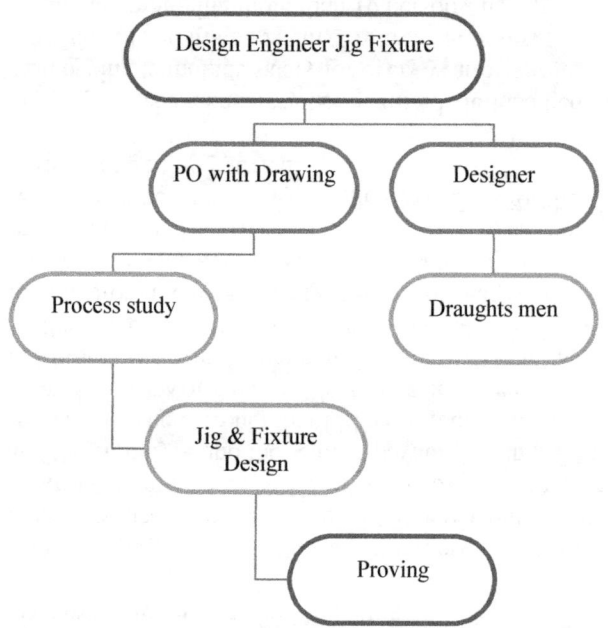

Designing of a jig and fixture is very easy, but while using it we have to prove it with cost effectiveness is very important.
In olden days, the Operators can do settings to do the jobs. For particular jobs, we can depend those Operators. There is no record for that. We can call that record as Jig and

As with most of the design fields, the idea for the design of a product arises from a need and has a use. Aesthetics is considered important in Product design but designers also deal with important aspects including technology, usability, human factors and material technology. Product designers are equipped with the skills needed to bring products from conception to market.

They should also have the ability to manage design projects, and subcontract areas to other sectors of the design industry. Example for product design is Gearboxes, Motor, Compressor, Vacuum cleaner, Herons, etc.

Design Engineer Jig and Fixture

Input for this engineer is the Purchase order and Drawing from sales. Once receiving those items first check the drawing number and revision number is matching or not. Then Study, the drawing, and write machining process with cross-functional team. What is cross-functional team? (CFT). The team, which contains the peoples from department head,'s like tool design, Production, Planning, Quality, Purchase, Sales, and stores. When ever the new product will come all will sit together and decide the process, Quality requirements, Tooling, Fixturing, time required for trial and dispatch of final product.

very big loss for that operator and as well as to the factory.

Factory HR again goes to search for a good operator will make expenses for advertising, time loss and production loss will be the result, up to that operator's recovery.

Safety equipments like Goggles, Hand glues, Shoes, Uniforms etc. should be provided.

Is it the uniform is the safety equipment? May be, give the uniform, and ask each operator to come with tuck in condition.

Because in the machine we cannot provide safeguard everywhere. For example, chuck and a lead screw in a lathe. The operator shirt will roll in to the lead screw in a lathe. Not only in the lathe, in every machine there is a loophole for rolling the shirt.

Recreation – Not only for the operators and for all the employees we have to give some refreshment. For example, provide small things like a TV in a lunchroom. Provide a small space to play a shuttle cork. Ask employees and their family to participate in Music, dance, Sports once in a year on Independence Day and give small prizes like gift articles and good books will give good result to the company.

Others – This facilities are not mandatory can be provided by big organizations or a good profit centers only. Those are Bonus, Incentives, Tour allowance, Sweets, and snacks at the time of festivals.

Therefore, before becoming as a supervisor we should know all the requirements provided by the management.

A Labor or an Operator comes often to the Supervisor for small grievances. Supervisor can fight with the management softly, recommend his practical skills, and help him. The owner or the management does not know about the operator's skill. The Supervisor can define that.

Environments

Working environment or the atmosphere is very important. For example, one Employer can say, we have given all the above facilities with better salary but still peoples resigning more. Why it is happening?

That is because of the working atmosphere. They will do painting, gas cutting, grinding, straightening, Rust proofing oil spray, and shouting everywhere inside the company. It has illiterate the operator.

Most of the company has started with out planning; they will start for their amount in their hand.

They will make a shed and thro' the machines and operators inside the shed. Make facilities one by one. They will not full fill the requirements immediately but give higher salary than the limited company and get failure in Labor management. Then permanently put a wanted board in his gate.

To day's trend, growth will be there but organization structure will not change drastically. So start a company as per ISO certification level, even its 2 machine or 3 machines. When company is, growing working atmosphere will also grow simultaneously and stop resignations.

We can now discuss about the Horizontal machining center is called HMC. The Horizontal boring machine is converted in to HMC machine.

The difference between the HMC and the horizontal boring machine are as follows.
The Lead screws are turned in to Ball screws.
The AC or DC motors are turned in to Servomotors.
Manual operations are computerized.
The ball screws will give greater accuracy than the lead screw and the servomotor will control the speeds well. We can take any CNC machines it will be built in this same manner only. If we are the person to innovate think alternate.
We can see the CNC machines are looking grand that is all if we remove the shirts of it, it will look like a 'Chicken'. Do not feels that design field is required more creative thinking and they will have a special mind. Every thing is possible by every body.

Here we discussed about the difference between two machines. However, I am not a Machine tool designer. In India, few of the Special purpose machine designer and manufacturer's are there. Mostly every body is making collaboration's and doing R&D. There we do not have any headaches.

Product Design - Product Design can be defined as the idea generation, concept development, testing and manufacturing or implementation of a physical object or service.

Cross feed of the Side carriage is functioning with the screw rod and get the drive from the Lead screw by disengaging the longitudinal movement of the side carriage. So far, we discussed in the mechanical side and in the electrical side is only a motor.

In this machine, Main Carriage movement we can take as X axis and Side carriage in which the tool post is mounted will move in Y direction. There is no Z axis movement.

So far, we discussed about the Conventional lathe. In this machine for X and Y movement machine feed is given, but will not move simultaneously. It will work individually and there is no Z movement. All over the world, still we are using this machine.

The new revolution is CNC machines. They provided the simultaneous machine feed for the X, Y, and Z directions with computerized control.

We do not think at any circumferences that, we did not have any brain compare to others. We do not compare with others life all ways try to stand our own legs.

All we know about the Seven Wonders of the World. How many times we will say splendid and marvelous. One, two, three or 10 times and get bored after that we will raise the questions, when, how and is it necessity to build such things now. Not necessary and we have created wonders through science now. The Flight, Rocket, Satellites and CNC machines.

Drawing study and Machines

Third step is drawing study. Now we are entering in to the pure technical side. The students may ask, Sir, for the above work itself more time is required and where will be the time for study the drawing.

In an Engineering company, drawing knowledge is more, first for the operator and next to the supervisor.

Every supervisor will have a training period for 2 to 3 years. In that training period, no body will trouble you. We are the free bird of the company. Those 3 years we have to get trained at least three or four machines. Should know the in and out of the machines. We will get close relation ship with the operators. Keenly watch their activities. What they are doing and how they are doing. This will help more and more when we will become as a Supervisor.

Turn over

Turn over and the productivity of the company is depends on a good Supervisor. The Turn over is fixed by the Sales and Planning department in Papers. That can be implemented and achieved by the Supervisor only.

A Supervisor is a person who will treat the employees in a nice manner will motivate them. A good Supervisor creates an environment where Operators thrive and bad ones either change or go out. Listen Operator complaint and suggestions.

If the Operator has a suggestion take the opportunity and utilize them rather than don't simply we can say we have tried that already it did not work. Complaints from the Operator about their assistants, helpers and other shifts to be listened carefully and take necessary action.

For Becoming as a Good Supervisor, listening is the good exercise. When the Operator coming with a complaints we call him in to the cabin and give respect and listen him in a sincere manner. So he will respect us and we do not know whether his problem will go to solve or not but ask him to give some time to solve it. In between, we can thing possibilities and does a root cause analysis.

A good Supervisor is like a good politician. Normally no body will accept the changes easily, it is a human nature. To become a good Supervisor be as a politician.

Shift Supervisor

After becoming a Good Supervisor, we are placed in shifts where the company is having night shifts. In day shifts distribution of loads, operator problems and more headaches will be there but it could be solved.

In night shifts, no Engineers or Managers will be available to solve the problems. The whole factory will come in to our control. In night shifts, we have to arrange medicines for fever, light injury's, Cough syrups, etc.

In addition, get details of a Van to get operator to the hospital if the injury is major. We are already trained in machines it will help to solve machining problems and small maintenance problems. The Supervisors are promoted to the next levels by their experiences.

The designer and drafter is usually directed by the design engineer. This stage is where design flaws are found and corrected and tooling, manufacturing fixtures, and packaging are developed.

We can take example as a Lathe. The following assemblies or products are available in a Conventional lathe.
- Headstock,
- Tailstock and a Carriage is a main particle.

The Headstock consists of Gears, change gears for threading engagement, pulley, motor, and the spindle. The motor drives pulley, from the pulley drive goes to gears and gears will control the speed of the spindle in which the chuck or faceplate is mounted. Now the headstock is functioning. In this machine, Headstock is a stationary item will not move in any direction.

The Tailstock consists of a Shaft with a taper seating in which we insert the revolving center or drills. The shaft itself will move back and forth by hand rotation. We are using it for drilling, tapping, and centering purpose only. The tailstock is moved by hand by loosening the hexagonal nut, which is clamped with the Lathe bed.

The main carriage consists of a Side carriage and a tool post in which tools are mounted. Its longitudinal movement is functioning with a lead screw rotation, which will get drive from the one end of head stock with gear engagement.

production. A model of the product is created and reviewed.

Prototypes are usually functional and non-functional. Functional prototypes are used for testing and the non-functional are used for form and fit checking.

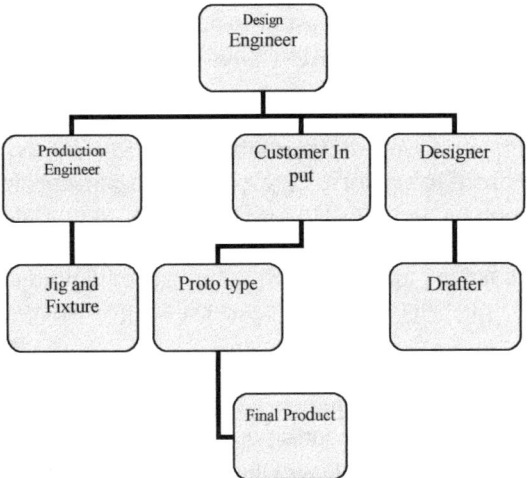

Final product Once the design is finalized the next step is preproduction. The design engineer, working with a manufacturing engineer and a quality engineer reviews an initial run of components and assemblies for design compliance. Variations in the product are correlated to aspects of the process and eliminated.

Production Engineer

The Supervisors from the machine shop are promoted to Production engineers by experience. The company will recruit the Degree holders directly as a Production Engineer instead of Supervisor.
They are also have a training period for two to three years. In that training period, they have to follow the same instructions given to the Supervisory grade. Once we will become as an Engineer get relaxed by following steps.

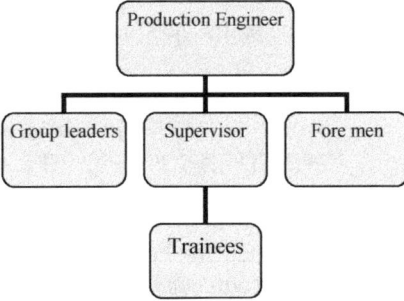

The production Engineer will get assistance like four or six Supervisors, Trainee peoples, fore men and Group leaders.

The Fore men and Group leaders promoted from the operator level by their experience and dedication to the company.

The Supervisor from seven to 8 years of his experience he will become an Engineer. Here his day-to-day activities are changing in to some other level.

He has to monitor his assistances and here the problems in every day machine shop activities and solve it. He is responsible to make benchmark details to his machine shop.

What are all the benchmark details? This will contain minimum to maximum Length, width, height, diameter, weight, surface finish, geometric tolerances of the work piece to be loaded in every machines. In addition, machine details like speed, feed, X.Y.Z. travels of the machine.

He does the cost cutting analysis. For cost cutting analysis we have already discussed in operator responsibility take that as a reference.

The production engineer has an experienced team now. Every problem to be discussed with their team members before solving it.

Once problem has solved is recorded and kept with the production engineer.

DESIGN ENGINEER

Design Engineer's minimum qualification is Bachelor of Engineering in Mechanical or Electrical. Machine tool design is simple and only we have to give drives and movements for the machine.

In olden days there is no software will be available for Design engineers, every thing will be done by practice only. Now there are many more software are available. For 2D drafting Auto cad is available and for 3D drafting soft wares like Pro-e, Uni graphics, Catia V5 and Solid Works is available. If the Machine shop is, the machine manufacturing company and they need the following Design engineers.

➢ Machine tool design Engineer
➢ Product Design engineer
➢ Jig and Fixture design Engineer.

Design engineer is classified as three types. In that, the Machine tool Design engineer is responsible for the design and development of new products, equipment, or facilities. Design engineer will get in put from the customers or for their own use get details from the shop floor. A Design Engineer should follow the following steps.

Customer In put introducing a product or for manufacturing a product, the customer input is essential.

Products are usually designed with input from a number of sources such as manufacturing, purchasing, and tool making engineering.

The next step in the design engineer's responsibility is prototyping. The design engineer has leads the project. The design engineer may direct a team of designers to create the CAD files necessary for prototyping and